Lecture Notes in Mathematics

Edited by A. Dold and B. Eckmann

1384

J. García-Cuerva (Ed.)

Harmonic Analysis and Partial Differential Equations

Proceedings of the International Conference held in El Escorial, Spain, June 9–13, 1987

Springer-Verlag

Berlin Heidelberg New York London Paris Tokyo Hong Kong

Editor

José García-Cuerva
Dpto. de Matemáticas C-XVI
Universidad Autónoma
28049 Madrid, Spain

Mathematics Subject Classification (1980): 42–xx, 35–xx, 30–xx

ISBN 3-540-51460-0 Springer-Verlag Berlin Heidelberg New York
ISBN 0-387-51460-0 Springer-Verlag New York Berlin Heidelberg

Printed in Germany

Printing and binding: Druckhaus Beltz, Hemsbach/Bergstr.
2146/3140-543210 – Printed on acid-free paper

These Proceedings are dedicated to the memory of José Luis Rubio de Francia (1949 - 1988)

INTRODUCTION

These are the Proceedings of the Conference on Harmonic Analysis and Partial Differential Equations, held in El Escorial, Spain, from June 9 to June 13, 1987. It was the third Conference of its kind in El Escorial; the other two took place in 1979 and 1983.

The program of the Conference included four main speakers who gave courses of three or four hours. The content of these courses is reflected in the four surveys or main lectures. Also included in these Proceedings are the ten 45-minute lectures which were of a more specialized nature.

Financial support for the Conference came from several sources: the Universidad Autónoma de Madrid, the CAICYT of the Spanish Government (Grant 2805/83) and the U.S.-Spanish Joint Committee for Scientific and Technological Cooperation (grant CCB-8402/058). We should like to thank CAJA MADRID for its support. We are also grateful to the town of El Escorial for its hospitality.

I shared the tasks of organizing the Conference with Professors Gene Fabes, Eugenio Hernández, José Luis Rubio de Francia and José Luis Torrea. Also a member of the Organizing Committee, and an essential one, was Caroline Bintcliffe, who acted as Secretary of the Conference.

We very much regret that José Luis Rubio de Francia is no longer with us. He died on February 6, 1988. He was one of the greatest mathematicians in our field as well as a wonderful person who was loved by all. It is only fitting that these Proceedings should be dedicated to his memory.

Let us hope that the tradition started in 1979 of holding an "El Escorial" every four years can be continued in 1991.

José García-Cuerva
Madrid, June 1988

Seminar on Harmonic Analysis and

Partial Differential Equations

9 - 13 June 1987. El Escorial (Spain)

LIST OF PARTICIPANTS

ADOLFSSON, Vilhelm (Chalmers U, Sweden)
AGUIRRE, Julián (UPV, Spain)
AIMAR, Hugo A. (INTEC, Argentina)
ALFARO, Manuel (U Zaragoza, Spain)
ALVAREZ, Josefina (Florida Atlantic U, USA)
ARIÑO, M.A. (U Barcelona, Spain)
BAERNSTEIN, Albert (Washington U, USA)
BALIBREA, Francisco (U Murcia, Spain)
BARCELO, Bartolomé (U Minnesota, USA)
BARCELO, Juan Antonio (UAM, Spain)
BARCELO, Miguel (ETSEIB, Spain)
BARQUILLA, Antonio (UAM, Spain)
BENEDETTO, John (U Maryland, USA)
BERNAL, Antonio (U Barcelona, Spain)
BLASCO, Oscar (U Zaragoza, Spain)
BONAMI, Aline (U Orléans, France)
BORJESON, Lennart (U Stockholm, Sweden)
BRUNA, Joaquim (UAB, Spain)
BURKHOLDER, Donald (U Illinois, USA)
CARBERY, Anthony (U Sussex, UK)
CARCANO, Giovanna (U Milano, Italy)
CARLSSON, Hasse (U Goteborg, Sweden)
CARRILLO, Ma. Teresa (UAM, Spain)
CARRO, Ma. Jesus (U Barcelona, Spain)
CASCANTE, Carmen (U Barcelona, Spain)
CERDA, Joan (U Barcelona, Spain)
CHRIST, Michael (UCLA, USA)
COBOS, Fernando (UAM, Spain)
COIFMAN, Ronald (Yale U, USA)
COLZANI, Leonardo (U Milano, Italy)
CORDOBA, Antonio (UAM, Spain)
DAVID, Guy (Ecole Polytechnique, France)
DINGER, Ulla (Chalmers U, Sweden)
DORRONSORO, José (UAM, Spain)
DROZDOWSKYJ, Ana (UAM, Spain)
DUOANDIKOETXEA, Javier (UAM, Spain)
ESCOBEDO, Miguel (UPV, Spain)
FABES, Eugene (U Minnesota, USA)
FEICHTINGER, Hans G. (U Wien, Austria)
FELDMAN, Mark (S Illinois U, USA)
FERNANDEZ-CABRERA, Luz M. (UAM, Spain)
GALLARDO, Diego (U Malaga, Spain)
GARCIA-CUERVA, José (UAM, Spain)
GAROFALO, Nicola (U Bologna, Italy)
GATTO, A. Eduardo (U Texas, USA)
GONZALEZ, Maria José (UAM, Spain)
GROCHENIG, Karlheinz (U Wien, Austria)
GUADALUPE, José (CU La Rioja, Spain)
GUIJARRO, Luis (UAM, Spain)
GUTIERREZ, Angel (UAM, Spain)
GUZMAN, Miguel de (UCM, Spain)
HEINIG, Hans P. (McMaster U, Canada)
HERNANDEZ, Eugenio (UAM, Spain)
HUGHES, Dale W. (Wichita State U, USA)
JANSON, Svante (Uppsala U, Sweden)
JAWERTH, Bjorn (Washington U, USA)
JOHNSON, Raymond L. (U Maryland, USA)
JONES, Peter (Yale U, USA)
KENIG, Carlos (U Chicago, USA)
KORENBLUM, Boris (State U New York, USA)
KUMLIN, Peter (Chalmers U, Sweden)
LOPEZ-HERRERO, Ma. Jesús (UAM, Spain)
MACIAS, Roberto (UAM, Spain)
MANFREDI, Juan J. (Purdue U, USA)
MARTIN, Miguel A. (UAM, Spain)
MARTIN, F. Javier (U Malaga, Spain)
MATEU, Joan (ETSIB, Spain)
MENARGUEZ, Trinidad (UAM, Spain)
MEYER, Yves (U Paris-Dauphine, France)
MILMAN, Mario (Florida Atlantic U, USA)
MIRA, José Manuel (U Murcia, Spain)
MOYUA, Adela (UPV, Spain)
NICOLAU, Artur (UAB, Spain)
OROBITG, Joan (UAB, Spain)
ORTEGA, Pedro (IB Malaga, Spain)
PABLO, Arturo de (UAM, Spain)
PASCUAS, Daniel (U Barcelona, Spain)
PERAL, Ireneo (UAM, Spain)
PEREZ, Mario (U Zaragoza, Spain)
REIMANN, Martin (U Bern, Switzerland)
RESINA, Ivam (U Campinas, Brasil)
REYES, Miguel (UAM, Spain)
REZOLA, Ma. Luisa (U Zaragoza, Spain)
RODRIGUEZ, José M. (UAM, Spain)
ROMERA, Elena (UAM, Spain)
RUBIO, José L. (UAM, Spain)
RUIZ, Alberto (UAM, Spain)
RUIZ, Francisco José (U Zaragoza, Spain)
SADOSKY, Cora (Howard U, USA)
SAWYER, Eric (McMaster U, Canada)
SEEGER, Andreas (TH Darmstadt, Germany)
SEMMES, Stephen (Yale U, USA)
SERAPIONI, Raul (U Trento, Italy)
SJOGREN, Peter (Chalmers U, Sweden)
SJOLIN, Per (Uppsala U, Sweden)
SORIA, Fernando (UAM, Spain)
SORIA, Javier (Washington U, USA)
SUEIRO, Juan (U Barcelona, Spain)
TAIBLESON, Mitchell (Washington U, USA)
TCHAMITCHIAN, Philippe (U Marseille, France)
TORREA, José L. (UAM, Spain)
TREBELS, Walter (TH Damstadt, Germany)
VAZQUEZ, Juan L. (UAM, Spain)
VEGA, Luis (UAM, Spain)
VERDERA, Joan (UAB, Spain)
VIVIANI, Beatriz (PEMA, Argentina)
WAINGER, Stephen (U Wisconsin, USA)
WALIAS, Magdalena (UAM, Spain)
WEISS, Guido (Washington U, USA)
WELLAND, Grant (U Missouri, USA)
WHEEDEN, Richard (Rutgers U, USA)
YABUTA, Kozo (Ibaraki U, Japan)
ZALOZNIK, Ales (U Ljubljana, Yugoslavia)

TABLE OF CONTENTS

DIFFERENTIAL SUBORDINATION OF HARMONIC FUNCTIONS AND MARTINGALES

Donald L. Burkholder

Department of Mathematics
University of Illinois
Urbana, Illinois 61801

A fruitful analogy in harmonic analysis is the analogy between a conjugate harmonic function and a martingale transform. One idea that underlies both of these concepts is the idea of differential subordination. Our study of it here yields new information about harmonic functions and martingales, and their interaction.

For example, let H be a real or complex Hilbert space with norm $|\cdot|$. Let u and v be harmonic in the open unit disk of $\mathbb{C}$ with values in H. If $|v(0)| \le |u(0)|$ and $|\nabla v(z)| \le |\nabla u(z)|$ for all z with $|z| < 1$, then

$$(0.1) \qquad m(\{\theta \in [0,2\pi) : |u(re^{i\theta})| + |v(re^{i\theta})| \ge 1\}) \le 2 \int_0^{2\pi} |u(re^{i\theta})|\,d\theta$$

where $0 < r < 1$ and m denotes Lebesgue measure. The constant 2 is best possible. The open unit disk can be replaced by a domain of $\mathbb{R}^n$ provided m is replaced by the harmonic measure of an appropriate subdomain. Inequality (0.1) is new even for $H = \mathbb{R}$.

Now suppose that $(\Omega, \mathcal{F}, \mu)$ is a positive measure space and $d_0, d_1, \ldots, e_0, e_1, \ldots$ are strongly integrable functions from Ω to H such that $|e_n(\omega)| \le |d_n(\omega)|$ for all $\omega \in \Omega$ and $n \ge 0$. Write $f_n = \sum_{k=0}^{n} d_k$ and $g_n = \sum_{k=0}^{n} e_k$. If, for all $n \ge 1$, both d_n and e_n are orthogonal to all bounded continuous H-valued functions of the pair f_{n-1}, g_{n-1}, then

$$(0.2) \qquad \mu(|f_n| + |g_n| \ge 1) \le 2 \int_\Omega |f_n|\,d\mu$$

and the constant 2 is best possible.

The proofs of (0.1) and (0.2) are similar and rest on the properties of the function $L: H \times H \to \mathbb{R}$ defined in (1.8). There are related maximal-function inequalities.

There are also analogous L^p-inequalities and their proofs rest on the function $U: H \times H \to \mathbb{R}$ defined in (2.5).

These inequalities have applications to square-function inequalities and to Riesz systems of functions satisfying the generalized Cauchy-Riemann equations. They yield extensions of the weak type (1,1) inequality of Kolmogorov [33] and the L^p-inequality of M. Riesz [41] for conjugate functions and the Hilbert transform. They also yield the analogues of these inequalities for martingale transforms [10]. They have, in addition, applications to martingale convergence and to the geometry of Banach spaces.

1. A WEAK-TYPE INEQUALITY FOR DIFFERENTIALLY SUBORDINATE HARMONIC FUNCTIONS

Let D be an open connected set of points $x = (x_1, \ldots, x_n) \in \mathbb{R}^n$ and H a real or complex Hilbert space with norm $|\cdot|$ and inner product $\langle \cdot, \cdot \rangle$. Suppose that $u: D \to H$ is harmonic: the partial derivatives $u_k = \partial u / \partial x_k$ and $u_{jk} = \partial^2 u / \partial x_j \partial x_k$ exist and are continuous, and $\Delta u = \sum_{k=1}^{n} u_{kk} = 0$, the origin of H. Note that $u_k(x) \in H$ and write

$$|\nabla u(x)| = \left(\sum_{k=1}^{n} |u_k(x)|^2\right)^{1/2} .$$

Suppose that $v: D \to H$ is also harmonic. Then v is <u>differentially</u> <u>subordinate</u> <u>to</u> u if, for all $x \in D$,

$$|\nabla v(x)| \leq |\nabla u(x)| .$$

Fix a point $\xi \in D$ and let D_0 be a bounded subdomain of D satisfying $\xi \in D_0 \subset D_0 \cup \partial D_0 \subset D$. Denote by $\mu = \mu_{D_0}^{\xi}$ the harmonic measure on ∂D_0 with respect to ξ. If $1 \leq p < \infty$, let

$$\|u\|_p = \sup_{D_0} \left[\int_{\partial D_0} |u|^p \, d\mu \right]^{1/p} \tag{1.1}$$

where the supremum is taken over all such D_0.

THEOREM 1.1. <u>Suppose that</u> u <u>and</u> v <u>are harmonic in the domain</u> $D \subset \mathbb{R}^n$ <u>with values in</u> H. <u>If</u> $|v(\xi)| \le |u(\xi)|$ <u>and</u> v <u>is differentially subordinate to</u> u, <u>then, with</u> $\mu = \mu_{D_0}^{\xi}$ <u>as above,</u>

$$\mu(|u| + |v| \ge 1) \le 2\|u\|_1 . \tag{1.2}$$

<u>The constant</u> 2 <u>is best possible.</u>

PROOF. Suppose there is a function $L: H \times H \to \mathbb{R}$ such that, for all s and t in H,

$$L(s,t) \le 2|s| \quad \text{if} \quad |s| + |t| \ge 1 , \tag{1.3}$$

$$L(s,t) \le 2|s| + 1 , \tag{1.4}$$

$$L(s,t) \ge 1 \quad \text{if} \quad |t| \le |s| , \tag{1.5}$$

$$L(u,v) \quad \text{is subharmonic on } D. \tag{1.6}$$

Then, by (1.3) and (1.4),

$$\begin{aligned} \mu(|u| + |v| \ge 1) &\le \mu(2|u| \ge L(u,v)) \\ &= \mu(2|u| - L(u,v) + 1 \ge 1) \\ &\le \int_{\partial D_0} [2|u| - L(u,v) + 1] d\mu . \end{aligned}$$

By (1.5) and (1.6),

$$\int_{\partial D_0} L(u,v) \ge L(u(\xi),v(\xi)) \ge 1 .$$

Therefore,

$$\mu(|u| + |v| \ge 1) \le 2 \int_{\partial D_0} |u| d\mu \le 2\|u\|_1 . \tag{1.7}$$

Equality holds, for example, if $H = \mathbb{R}$ and $u = v \equiv 1/2$. (Also, see Remark 1.1.)

The question is: Does there exist a function $L: H \times H \to \mathbb{R}$ with the properties (1.3) through (1.6)?

Consider the function L given by

$$(1.8)\qquad L(s,t) = 1 + |s|^2 - |t|^2 \quad \text{if} \quad |s| + |t| < 1 \,,$$
$$= 2|s| \quad \text{if} \quad |s| + |t| \geq 1 \,.$$

Property (1.3) is satisfied and this implies that (1.4) is satisfied under the condition $|s| + |t| \geq 1$. But if $|s| + |t| < 1$, then $|s|^2 < |s| < 1$ and $1 + |s|^2 - |t|^2 \leq 1 + 2|s|$ so (1.4) is satisfied here also. To prove (1.5), let $|t| \leq |s|$. Then the condition $|s| + |t| < 1$ implies that $L(s,t) = 1 + |s|^2 - |t|^2 \geq 1$ and the condition $|s| + |t| \geq 1$ implies that $L(s,t) = 2|s| \geq |s| + |t| \geq 1$. To prove (1.6), let W and V be defined on $H \times H$ as follows:

$$W(s,t) = 1 + |s|^2 - |t|^2 \,,$$
$$V(s,t) = 2|s| \,.$$

Then $L(s,t) = W(s,t) > V(s,t)$ if $|s| + |t| < 1$ and $L(s,t) = V(s,t)$ otherwise. The condition $|s| + |t| \leq 1$ implies that $1 - |s| \geq |t|$ which in turn implies that $W(s,t) = 1 + |s|^2 - |t|^2 \geq 2|s| = V(s,t)$, with equality holding if and only if $|s| + |t| = 1$.

We can now show that the continuous function $L(u,v)$ is subharmonic in D by showing that

$$(1.9)\qquad A(L(u,v),x,r) \geq L(u(x),v(x))$$

for all small positive r and all $x \in D$, where the expression on the left-hand side denotes the average of $L(u,v)$ over the ball with radius r and center x. Note that

$$\begin{aligned} A(V(u,v),x,r) &= 2A(|u|,x,r) \\ &\geq 2|A(u,x,r)| \\ &= 2|u(x)| = V(u(x),v(x)) \,, \end{aligned}$$

so $V(u,v)$ is subharmonic in D. The function $W(u,v)$ is also subharmonic in D because

$$\Delta W(u,v) = 2(|\nabla u|^2 - |\nabla v|^2) \geq 0 \,.$$

Therefore (1.9) holds for all small r and all $x \in D$ satisfying $|u(x)| + |v(x)| \neq 1$. If $x \in D$ and $|u(x)| + |v(x)| = 1$, then (1.9) must hold for all small r because

$$A(L(u,v),x,r) \geq A(V(u,v),x,r)$$
$$\geq V(u(x),v(x)) = L(u(x),v(x)) .$$

This completes the proof of Theorem 1.1. Note that (0.1) is a special case of (1.7).

REMARK 1.1. If $f \in L^1_H([0,2\pi))$ and $\tilde{H}f$ is its periodic Hilbert transform, then

$$(1.10) \qquad m(|f| + |\tilde{H}f| \geq 1) \leq 2\|f\|_1 .$$

This is an immediate consequence of (0.1) and the boundary limit theorems of Fatou and Privalov in this setting. The constant 2 is best possible. To see this, assume $H = \mathbb{R}$ and let F be an analytic univalent function that maps the open unit disk of $\mathbb{C}$ onto $\{w \in \mathbb{C}: |\mathrm{Re}\, w| + |\mathrm{Im}\, w| < 1\}$ with $F(0) = 0$. Let $f + i\tilde{H}f$ be the almost everywhere radial limit of F. Then

$$m(|f| + |\tilde{H}f| = 1) = 2\pi = \|f\|_1 + \|\tilde{H}f\|_1 .$$

Since $\tilde{H}^2 = -\tilde{H}$, inequality (1.10) applied to $\tilde{H}f$ gives $\|\tilde{H}f\|_1 \geq \pi$ so $\|f\|_1 \leq \pi$ and equality holds in (1.10).

REMARK 1.2. If $f \in L^1_H(\mathbb{R})$ and Hf is its Hilbert transform, then

$$(1.11) \qquad m(|f| + |Hf| \geq 1) \leq 2\|f\|_1 .$$

This follows from (1.10) with the use of a classical argument [50]. Again the constant 2 is best possible. This can be seen by modifying an example of Davis [23] as follows. Consider the mapping

$$w \mapsto -(w + w^{-1})/2$$

from the upper half of the open unit disk onto the upper half-plane. Let G be its inverse. On the boundary $\mathbb{R}$, the value of G at x is unimodular if $|x| \leq 1$ and satisfies $|G(x)| \leq 1/|x|$ if $|x| \geq 1$. Let F be the conformal mapping of Remark 1.1. By the Schwarz lemma, $|F(w)| \leq |w|$ if $|w| < 1$, so $|F([G(z)]^n)| \leq |G(z)|^n$, which converges to 0 as $z \to \infty$ in the upper half-plane.

Let f_n be the real part of $F(G^n)$ on $\mathbb{R}$ and Hf_n the imaginary part on $\mathbb{R}$. Then, for $n \geq 2$,

$$m(|f_n| + |Hf_n| = 1) = m([-1,1]) = 2$$

and

$$\|f_n\|_1 + \|Hf_n\|_1 \leq \int_{-1}^{1} dx + 4 \int_{1}^{\infty} x^{-n}\, dx$$

$$= 2 + 4/(n-1) .$$

Using (1.11) and $H^2 = -H$, we have both $\|f_n\|_1 \geq 1$ and $\|Hf_n\|_1 \geq 1$, so that $\|f_n\|_1 \leq 1 + 4/(n-1)$. Letting $n \to \infty$, we see that 2 is best possible.

REMARK 1.3. Under the conditions of Theorem 1.1, the inequality

$$\mu(|v| \geq 1) \leq c\|u\|_1 \tag{1.12}$$

follows at once from (1.2) with $c = 2$. The best value of c is not yet known. Davis [23] discovered the value of the best constant K in Kolmogorov's inequality [33]: If $f \in L^1_{\mathbb{R}}([0,2\pi))$, then

$$m(|\tilde{H}f| \geq 1) \leq K\|f\|_1 . \tag{1.13}$$

It is given by

$$K = \frac{1 + 3^{-2} + 5^{-2} + 7^{-2} + \dots}{1 - 3^{-2} + 5^{-2} - 7^{-2} + \dots} .$$

Therefore, $K \leq c \leq 2$. (see Baernstein [1] for another derivation of K.)

As Baernstein has observed (see [39]), the best value of the constant in (1.13) is not known if $\mathbb{R}$ is replaced by $\mathbb{C}$. By (1.12), it cannot exceed 2. In the analogue for martingale transforms, it is 2 (see (3.26) in [12]).

The following inequality for a natural maximal function associated with u and v extends (1.2). Let $(\Omega,\mathcal{F},P)$ be a probability space, $X = (X_t)_{t \geq 0}$ a Brownian motion in $\mathbb{R}^n$ starting at $\xi \in D \subset \mathbb{R}^n$, and T_D the first time that X leaves D: for $\omega \in \Omega$,

$$T_D(\omega) = \inf\{t > 0: X_t(\omega) \notin D\} .$$

THEOREM 1.2. <u>Suppose that</u> u <u>and</u> v <u>are harmonic in the domain</u> $D \subset \mathbb{R}^n$ <u>with values in</u> H. <u>If</u> $|v(\xi)| \leq |u(\xi)|$ <u>and</u> v <u>is differentially subordinate to</u>

u, <u>then</u>

(1.14) $$P(\sup_{t < T_D} [|u(X_t)| + |v(X_t)|] \geq 1) \leq 2\|u\|_1 .$$

Here $\|u\|_1$ has the same meaning as before or, equivalently, is given by $\sup_{D_0} E|u(X_{T_0})|$ where $T_0 = T_{D_0}$. The constant 2 is again best possible since it is already best possible in (1.2).

PROOF. Let $T = \inf\{t > 0: |u(X_{t \wedge T_0})| + |v(X_{t \wedge T_0})| > 1\}$ and $S = T \wedge T_0$. Then, because D_0 is bounded, T_0 is finite almost everywhere and

(1.15) $$P(\sup_{t < T_0} [|u(X_t)| + |v(X_t)|] > 1) \leq P(|u(X_S)| + |v(X_S)| \geq 1) .$$

By the same reasoning as in the proof of Theorem 1.1, the right-hand side is no greater than

$$E[2|u(X_S)| - L(u(X_S),v(X_S)) + 1] \leq 2E|u(X_S)| \leq 2E|u(X_{T_0})| .$$

Here we have used (1.5) and (1.6). Note that (1.6) implies that $(L(u(X_{t \wedge S}),v(X_{t \wedge S})))_{t \geq 0}$ is a submartingale [25] and hence is expectation-nondecreasing.

Letting $D_0 \uparrow D$, we obtain (1.14) with "$\geq$" replaced by "$>$". This implies the stronger inequality and completes the proof.

REMARK 1.4. If $\|u\|_1$ is finite, then Theorem 1.2 implies that, with probability one, the Brownian maximal function of u and the Brownian maximal function of v are finite. Therefore, for almost all $\omega \in \Omega$, the limits

$$\lim_{t \uparrow T_D(\omega)} u(X_t(\omega)) \quad \text{and} \quad \lim_{t \uparrow T_D(\omega)} v(X_t(\omega))$$

exist and are finite. See Doob [26] and the references given there.

2. AN L^p-INEQUALITY FOR DIFFERENTIALLY SUBORDINATE HARMONIC FUNCTIONS

THEOREM 2.1. <u>Let</u> $1 < p < \infty$ <u>and</u> p^* <u>be the maximum of</u> p <u>and</u> q <u>where</u> $1/p + 1/q = 1$. <u>Suppose that</u> u <u>and</u> v <u>are harmonic in the domain</u> $D \subset \mathbb{R}^n$ <u>with values in</u> H. <u>If</u> $|v(\xi)| \leq |u(\xi)|$ <u>for some</u> $\xi \in D$ <u>and</u> v <u>is differentially</u>

subordinate to u, *then*

(2.1) $$\|v\|_p \le (p^* - 1)\|u\|_p \ .$$

Recall that H is any real or complex Hilbert space and $\|u\|_p$, which depends on ξ, is defined by (1.1).

PROOF. Let D_0 be a bounded subdomain of D satisfying $\xi \in D_0 \subset D_0 \cup \partial D_0 \subset D$, and let $\mu = \mu_{D_0}^{\xi}$ be the harmonic measure on ∂D_0 with respect to ξ. Suppose there is a function $U\colon H \times H \to \mathbb{R}$ such that, for all s and t in H,

(2.2) $$U(s,t) \ge |t|^p - (p^* - 1)^p |s|^p \ ,$$

(2.3) $$U(s,t) \le 0 \quad \text{if} \quad |t| \le |s| \ ,$$

(2.4) $$U(u,v) \quad \text{is superharmonic on} \quad D.$$

Then, by (2.2), $|v|^p \le (p^* - 1)^p |u|^p + U(u,v)$, so that

$$\int_{\partial D_0} |v|^p \, d\mu \le (p^* - 1)^p \int_{\partial D_0} |u|^p \, d\mu + \int_{\partial D_0} U(u,v) d\mu \ .$$

By (2.3) and (2.4),

$$\int_{\partial D_0} U(u,v) d\mu \le U(u(\xi), v(\xi)) \le 0 \ .$$

Therefore inequality (2.1) follows from the existence of such a function U, the existence of which we now demonstrate.

Let $U\colon H \times H \to \mathbb{R}$ be defined by

(2.5) $$U(s,t) = \alpha_p(|t| - (p^* - 1)|s|)(|s| + |t|)^{p-1}$$

where $\alpha_p = p(1 - 1/p^*)^{p-1}$. This function plays an essential role in [18] where (2.2) and (2.3), together with other properties, are proved. Also, see [19]. It remains to show (2.4).

Define M and N on D by $M(x) = |u(x)|$ and $N(x) = |v(x)|$. Assume temporarily that $u(x) \ne 0$ and $v(x) \ne 0$ for all $x \in D$. Then M and N are infinitely differentiable on D. If $p \ge 2$, then

(2.6) $$\Delta U(u,v) = -\alpha_p(A + B + C)$$

where

$$A = p(p-1)(|\nabla u|^2 - |\nabla v|^2)(M + N)^{p-2} \geq 0 ,$$

$$B = p(p-2)(\Delta N)(M + N)^{p-1} \geq 0 ,$$

$$C = p(p-1)(p-2)|\nabla(M + N)|^2 M(M + N)^{p-3} \geq 0 .$$

If $1 < p \leq 2$, then

$$(p-1)\Delta U(u,v) = \alpha_p(A^0 + B^0 + C^0)$$

where the expressions for A^0, B^0, C^0 are the same as those for A, B, C with u and v (and M and N) interchanged; by the identity $(p-1)(p^* - 1) = 1$, which is valid for $1 < p \leq 2$,

$$(p-1)U(s,t) = -\alpha_p(|s| - (p-1)|t|)(|t| + |s|)^{p-1} .$$

Therefore, if $1 < p < \infty$, then $\Delta U(u,v) \leq 0$ and $U(u,v)$ is superharmonic, at least under the condition that u and v do not vanish.

We can assume that u and v take their values in a proper closed subspace H_0 of H (otherwise enlarge H slightly). Let $a \neq 0$ be in the orthogonal complement of H_0. Then the harmonic functions $a + u$ and $a + v$ satisfy the assumptions of the theorem. In addition, $|a + u(x)| \geq |a| > 0$ with a similar result for v. By what we have already proved, $U(a + u,a + v)$ is superharmonic. Letting $a \to 0$, we obtain (2.4) and hence the inequality (2.1).

REMARK 2.1. Theorem 2.1 extends the M. Riesz inequality [41] in which H is the real line, D is the open unit disk, and v is conjugate to u with $v(0) = 0$. Pichorides [40] and Cole (see [29]) have obtained the best constant in the Riesz inequality: $\cot(\pi/2p^*)$. (Also, see Pełczyński [39].) The best constant in the inequality $\|v\|_p \leq c_p\|u\|_p$, where v is differentially subordinate to u, and the other conditions of Theorem 2.1 hold, must satisfy $\cot(\pi/2p^*) \leq c_p \leq p^* - 1$. In each of the two analogous inequalities, one for martingale transforms and the other for differentially subordinate martingales, the best constant is $p^* - 1$ (see [16] and [19]).

3. AN APPLICATION TO RIESZ SYSTEMS

Let $u_0, u_1, \ldots, u_n$ be harmonic in a domain D of points $x = (x_0, \ldots, x_n)$ of $\mathbb{R}^{n+1}$. Assume that the u_k have their values in a real or complex Hilbert space H and that the generalized Cauchy-Riemann equations are satisfied:

$$\text{(3.1)} \qquad \sum_{k=0}^{n} u_{kk} = 0 \quad \text{and} \quad u_{jk} = u_{kj} ,$$

where $u_{jk} = \partial u_j / \partial x_k$, for $j,k = 0,\ldots,n$. Recall that if $u : D \to H$ is harmonic, then $u_0 = \partial u/\partial x_0, \ldots, u_n = \partial u/\partial x_n$ satisfy these equations. These systems of harmonic functions can be used to extend the classical theory of Hardy spaces as Stein and Weiss [47] have shown.

Let $F = (u_0, \ldots, u_n)$, $F^{00} = (u_0, 0, \ldots, 0)$, and $F^0 = (0, u_1, \ldots, u_n)$. These are harmonic functions from D to K where $K = H \times \cdots \times H$ and the norm of $(s_0, \ldots, s_n) \in K$ is given by $(\sum_{j=0}^{n} |s_j|^2)^{1/2}$. It is easy to see that F^{00} is differentially subordinate to $n^{1/2}F^0$ and that F is differentially subordinate to $(n+1)^{1/2}F^0$. For example, $|\nabla F^{00}| = |\nabla u_0|$ where, by (3.1),

$$\begin{aligned}
|\nabla u_0|^2 &= |u_{00}|^2 + \sum_{k=1}^{n} |u_{0k}|^2 \\
&= \Big|\sum_{k=1}^{n} u_{kk}\Big|^2 + \sum_{k=1}^{n} |u_{k0}|^2 \\
&\le (n-1) \sum_{k=1}^{n} |u_{kk}|^2 + \Big(\sum_{k=1}^{n} |u_{kk}|^2 + \sum_{k=1}^{n} |u_{k0}|^2\Big) \\
&\le (n-1)|\nabla F^0|^2 + |\nabla F^0|^2 \\
&= n|\nabla F^0|^2 .
\end{aligned}$$

The constant n is best possible: consider the gradient of $nx_0^2 - x_1^2 - \cdots - x_n^2$.

Therefore, by Theorems 1.1 and 2.1, if there is a point $\xi \in D$ such that $|u_0(\xi)| \le n^{1/2}|F^0(\xi)|$ and μ has the same meaning as before, then

$$\text{(3.2)} \qquad \mu(|u_0| \ge 1) \le 2n^{1/2}\|F^0\|_1$$

and, for $1 < p < \infty$,

$$\text{(3.3)} \qquad \|u_0\|_p \le n^{1/2}(p^* - 1)\|F^0\|_p .$$

There are similar inequalities between F and F^{U}. See Stein [46] and Essén [27] for related results.

4. SOME ANALOGOUS INEQUALITIES

Let Ω be a nonempty set, H a real or complex Hilbert space as before, and $f = (f_n)_{n \geq 0}$ a sequence of functions from Ω to H. Another such sequence $g = (g_n)_{n \geq 0}$ is <u>differentially subordinate</u> to f if, for all $\omega \in \Omega$ and $k \geq 0$,

$$(4.1) \qquad |e_k(\omega)| \leq |d_k(\omega)|$$

where, for all $n \geq 0$,

$$f_n = \sum_{k=0}^{n} d_k \quad \text{and} \quad g_n = \sum_{k=0}^{n} e_k .$$

Here let μ be any positive measure on some σ-field of subsets of Ω. Suppose that $f_0, f_1, \ldots, g_0, g_1, \ldots$ are strongly measurable and set $\|f\|_p = \sup_{n \geq 0} \|f_n\|_p$. If $s,t \in H$, let (s,t) denote the real part of $\langle s,t \rangle$.

THEOREM 4.1. <u>Suppose that</u> g <u>is differentially subordinate to</u> f <u>as above. If</u> $\|f\|_1$ <u>is finite and</u>

$$(4.2) \qquad \int_\Omega (\varphi(f_{n-1},g_{n-1}),d_n)\,d\mu = \int_\Omega (\varphi(f_{n-1},g_{n-1}),e_n)\,d\mu = 0$$

<u>for all bounded continuous functions</u> φ <u>from</u> $H \times H$ <u>to</u> H <u>and all positive integers</u> n, <u>then</u>

$$(4.3) \qquad \mu(|f_n| + |g_n| \geq 1) \leq 2\|f\|_1 .$$

<u>The constant</u> 2 <u>is best possible</u>.

Here $H \times H$ is given its natural Hilbert-space norm and "best possible" is to mean that if $\beta < 2$, then there exist f and g as above on some measure space (any nonatomic space would do) such that the left-hand side of (4.3) exceeds $\beta\|f\|_1$. Actually, the constant 2 is already best possible in the weaker inequality $\mu(|g_n| \geq 1) \leq 2\|f\|_1$ and equality can hold. See [11] or consider the following example. Let $\Omega = \{-1/2, 3/2\}$ with $\mu(\{-1/2\}) = 3/4$ and $\mu(\{3/2\}) = 1/4$.

Let $e_0 = d_0 \equiv 1/2$ and $e_1 = -d_1$ where $d_1(\omega) = \omega$ for $\omega \in \Omega$. Finally, let $e_n = d_n \equiv 0$, $n \geq 2$. Then the assumptions of the theorem are satisfied and, for $n \geq 1$, we have $|g_n| \equiv 1$ so that $\mu(|g_n| \geq 1) = 1$ while $\|f\|_1 = \|f_1\|_1 = 0 \cdot (3/4) + 2 \cdot (1/4) = 1/2$.

The f and g of Theorem 4.1 are not necessarily martingales but they are very weak martingales (cf. Section 13 of [16]).

PROOF. Let $L: H \times H \to H$ be the function defined in (1.8). Then

$$\mu(|f_n| + |g_n| \geq 1) \leq \int_\Omega [2|f_n| - L(f_n,g_n) + 1]d\mu$$

by the same reasoning used in the proof of Theorem 1.1. If

$$\int_\Omega [L(f_n,g_n) - 1]d\mu \geq 0 , \tag{4.4}$$

then $\mu(|f_n| + |g_n| \geq 1) \leq 2\|f_n\|_1$ and (4.3) holds.

To prove (4.4), we shall use the functions $\varphi: H \times H \to H$ and $\psi: H \times H \to H$ defined as follows:

$$\begin{aligned} \varphi(s,t) &= 2s \quad \text{if} \quad |s| + |t| < 1 \quad \text{or} \quad s = 0 , \\ &= 2s/|s| \quad \text{if} \quad |s| + |t| \geq 1 \quad \text{and} \quad s \neq 0 , \end{aligned}$$

$$\begin{aligned} \psi(s,t) &= -2t \quad \text{if} \quad |s| + |t| < 1 , \\ &= 0 \quad \text{if} \quad |s| + |t| \geq 1 . \end{aligned}$$

We shall show that these functions satisfy the inequality

$$L(s+h,t+k) \geq L(s,t) + (\varphi(s,t),h) + (\psi(s,t),k) \tag{4.5}$$

for all $s,t,h,k \in H$ such that $|k| \leq |h|$. Therefore, by (4.1),

$$\begin{aligned} L(f_0,g_0) - 1 &= L(d_0,e_0) - L(0,0) \\ &\geq (\varphi(0,0),d_0) + (\psi(0,0),e_0) = 0 , \end{aligned}$$

which also follows from (1.5), and, for $n \geq 1$,

$$L(f_n,g_n) \geq L(f_{n-1},g_{n-1}) + (\varphi(f_{n-1},g_{n-1}),d_n) + (\psi(f_{n-1},g_{n-1}),e_n) \tag{4.6}$$

Accordingly, $\int_\Omega [L(f_0,g_0) - 1]d\mu \geq 0$ and the monotonicity property

(4.7) $$\int_\Omega [L(f_n,g_n) - 1]d\mu \geq \int_\Omega [L(f_{n-1},g_{n-1}) - 1]d\mu$$

will follow from

(4.8) $$\int_\Omega (\varphi(f_{n-1},g_{n-1}),d_n)d\mu = 0$$

and a similar equality for the last term of (4.6). To see that (4.8) holds, construct a sequence of continuous functions φ_j from $H \times H$ to H such that $|\varphi_j(s,t)| \leq 2$ and $\lim_{j\to\infty} \varphi_j(s,t) = \varphi(s,t)$ for all $s,t \in H$. The function $(\varphi_j(f_{n-1},g_{n-1}),d_n)$ is measurable and is bounded in absolute value by the integrable function $2|d_n|$. By (4.2) applied to φ_j and the dominated convergence theorem, (4.8) holds.

Fix $s,t,h,k \in H$ with $|k| \leq |h|$. To see that (4.5) holds, consider $G: \mathbb{R} \to \mathbb{R}$ defined by

$$G(\alpha) = L(s + \alpha h, t + \alpha k) .$$

By following the pattern of the proof of (1.6), observe that G is locally convex. Therefore G is convex on $\mathbb{R}$. Accordingly,

(4.9) $$G(1) \geq G(0) + G'(0)$$

provided $G'(0)$ exists. To see that (4.5) follows, we consider four cases. Case (i): $|s| + |t| < 1$. In this case, $G'(0) = (\varphi(s,t),h) + (\psi(s,t),k)$ and (4.9) gives (4.5). Case (ii): $|s| + |t| > 1$ and $s \neq 0$. Similar to (i). Case (iii): $|s| + |t| \geq 1$ and $s \neq 0$. Let $s_j \in H$, $|s_j| > |s|$, and $\lim_{j\to\infty} s_j = s$. By (ii), the desired inequality (4.5) holds if s is replaced by s_j. Take limits of both sides as $j \to \infty$. Case (iv): $|s| + |t| \geq 1$, $s = 0$, so $|t| \geq 1$. Here the right-hand side of (4.5) is $2|s| + (0,h) + (0,k) = 0$. If $|s + \alpha h| + |t + \alpha k| \geq 1$, then the left-hand side of (4.5) is $2|s + \alpha h| \geq 0$. If $|s + \alpha h| + |t + \alpha k| < 1$, then $|t + \alpha k| < 1$ and the left-hand side exceeds $|s + \alpha h|^2 \geq 0$. This completes the proof of (4.5), (4.4), and (4.3).

THEOREM 4.2. _Let_ $1 < p < \infty$ _and_ p^* _be the maximum of_ p _and_ q _where_ $1/p + 1/q = 1$. _Suppose that_ $\|f\|_p$ _is finite and_ g _is differentially subordinate to_ f. _If_ $n \geq 1$, _let_ (4.2) _hold for all continuous functions_ φ _from_ $H \times H$ _to_

H *satisfying* $\|\varphi(f_{n-1},g_{n-1})\|_q < \infty$. *Then*

(4.10) $$\|g\|_p \le (p^* - 1)\|f\|_p .$$

The constant $p^* - 1$ *is best possible. If* $\|f\|_p$ *is both positive and finite, then equality holds in* (4.10) *if and only if* $p = 2$ *and equality holds in* (4.1) *for almost all* ω *and all* $k \ge 0$.

The proof is the same as the proof of Theorem 1.1 in [19], where it is assumed that $\mu(\Omega) = 1$ and that f and g are martingales relative to the same increasing sequence of sub-σ-fields.

REMARK 4.1. Let $1 < p < \infty$ and suppose that $h = (h_n)_{n \ge 0}$ is the sequence of Haar functions on the Lebesgue unit interval. The following inequality from [18] is an immediate consequence of (4.10). If $a_k, b_k \in H$ and $|b_k| \le |a_k|$, then for all $n \ge 0$,

(4.11) $$\|\sum_{k=0}^{n} b_k h_k\|_p \le (p^* - 1)\|\sum_{k=0}^{n} a_k h_k\|_p .$$

If $H = \mathbb{R}$ and $b_k = \varepsilon_k a_k$ where $\varepsilon_k \in \{1,-1\}$, the constant $p^* - 1$ is already best possible ([14] and [16]). As was noted in [18], Pełczyński's conjecture [39] follows: The inequality of Paley [38] and Marcinkiewicz [35] for the Haar system holds with the same constant if the multiplier sequence of signs ± 1 is replaced by a sequence of unimodular complex numbers.

REMARK 4.2. Let μ be a positive measure and $d = (d_n)_{n \ge 0}$ a sequence of strongly integrable functions from Ω to H such that

(4.12) $$\int_\Omega (\varphi(d_0,\dots,d_{n-1}),d_n)\,d\mu = 0$$

for all bounded continuous functions φ from $H \times \cdots \times H$ to H and all positive integers n. Let $S_n(f) = (\sum_{k=0}^{n} |d_k|^2)^{1/2}$ where f is the sequence of partial sums of d as before. Then, for example,

(4.13) $$\mu(|f_n| + S_n(f) \ge 1) \le 2\|f\|_1 .$$

The constant 2 is best possible: consider $d_0 \equiv 1/2$ and $d_n = 0$, $n \ge 1$, on a probability space. Inequality (4.13) follows from Theorem 4.1 applied to $F = (F_n)_{n \geq 0}$ and $G = (G_n)_{n \geq 0}$ with difference sequences D and E respectively:

$$D_k(\omega) = (d_k(\omega),0,0,\ldots),$$
$$E_k(\omega) = (0,\ldots,0,d_k(\omega),0,\ldots),$$

so that $F_n = (f_n,0,0,\ldots)$ and $G_n = (d_0,\ldots,d_n,0,0,\ldots)$. These are K-valued where $K = \ell^2_H$, the Hilbert space of sequences $s = (s_0,s_1,\ldots)$ with $s_j \in H$ and

$$|s| = (\sum_{j=0}^{\infty} |s_j|^2)^{1/2} < \infty .$$

The orthogonality assumption (4.2) is satisfied and $|E_k(\omega)|_K = |D_k(\omega)|_K$. Furthermore, $|F_n|_K = |f_n|$ and $|G_n|_K = S_n(f)$. Therefore, (4.13) follows from (4.3).

Similarly, if $1 < p < \infty$ and the d_n satisfy $\|d_n\|_p < \infty$ and the orthogonality assumption (4.12) for all continuous functions φ from $H \times \cdots \times H$ to H such that $\|\varphi(d_0,\ldots,d_{n-1})\|_q$ is finite, then

$$(4.14) \qquad (p^* - 1)^{-1}\|S(f)\|_p \leq \|f\|_p \leq (p^* - 1)\|S(f)\|_p$$

where $S(f) = (\sum_{k=0}^{\infty} |d_k|^2)^{1/2}$ and $\|f\|_p = \sup_n \|f_n\|_p$ as before. See [19] for further discussion and applications to stochastic integrals. Also, see Theorem 5.2 below and the inequalities for $S(g,T)$ in (5.1) and (5.2).

REMARK 4.3. The assumptions of Theorem 4.1 are too weak to give the maximal inequality

$$(4.15) \qquad \mu(\sup_{n \geq 0} [|f_n| + |g_n|] \geq 1) \leq 2\|f\|_1 .$$

See the example, for the case $H = \mathbb{R}$ and $\mu(\Omega) = 1$, described on page 696 of [16].

However, suppose that (4.2) is replaced by the stronger assumption that

$$(4.16) \qquad \int_\Omega (\varphi(d_0,e_0,\ldots,d_{n-1},e_{n-1}),d_n)d\mu = 0$$

and

$$(4.17) \qquad \int_\Omega (\varphi(d_0,e_0,\ldots,d_{n-1},e_{n-1}),e_n)d\mu = 0$$

for all bounded continuous functions φ from $H \times \cdots \times H$ to H. Then (4.15) does hold. To see this, let $w_0 \equiv 1$ and, for $n \geq 1$, let $w_n(\omega) = 1$ if $\sup_{k<n} [|f_k(\omega)| + |g_k(\omega)|] \leq 1$ and $w_n(\omega) = 0$ otherwise. Then $F = (F_n)_{n \geq 0}$ and $G = (G_n)_{n \geq 0}$ defined by $F_n = \sum_{k=0}^{n} w_k d_k$ and $G_n = \sum_{k=0}^{n} w_k e_k$ satisfy the assumptions of Theorem 4.1, so that

$$\mu(\sup_{0 \leq k \leq n} [|f_k| + |g_k|] > 1) \leq \mu(|F_n| + |G_n| > 1)$$
$$\leq 2\|F_n\|_1 \leq 2\|f\|_1 .$$

This implies (4.15). On the other hand, if $g^*(\omega) = \sup_{n \geq 0} |g_n(\omega)|$, then it is not always true that $\mu(f^* + g^* \geq 1) \leq 2\|f\|_1$. (Let $F = \alpha f$ and $G = \alpha g$ where $\alpha = 2/3$ and f and g are defined as in the example just after the statement of Theorem 4.1.) Inequality (4.15) does imply that

$$\mu(g^* \geq 1) \leq 2\|f\|_1 . \tag{4.18}$$

5. SOME APPLICATIONS TO MARTINGALES

Let $(\Omega, \mathfrak{F}_\infty, P)$ be a probability space and $\mathfrak{F} = (\mathfrak{F}_n)_{n \geq 0}$ a nondecreasing sequence of sub-σ-fields of $\mathfrak{F}_\infty$. Suppose that B is a real or complex Banach space with norm $|\cdot|$, and $f = (f_n)_{n \geq 0}$ and $g = (g_n)_{n \geq 0}$ are B-valued martingales. That is, d_n and e_n are strongly $\mathfrak{F}_n$-measurable and integrable functions from Ω to B with $E(d_{n+1}|\mathfrak{F}_n) = E(e_{n+1}|\mathfrak{F}_n) = 0$ for all $n \geq 0$.

THEOREM 5.1. *Suppose that B is isomorphic to a Hilbert space. If $\|f\|_1$ is finite and g is differentially subordinate to f, then g converges almost everywhere.*

For the case $B = \mathbb{R}$, see [10]. The proof of the theorem is new even for the special case $g = f$ and $B = \mathbb{R}$, which is Doob's theorem [24]: If f is a real-valued martingale with $\|f\|_1$ finite, then f converges almost everywhere.

PROOF. We can assume that $B = H$. Let $\varepsilon > 0$ and

$$A_\varepsilon = \{\omega \in \Omega: \inf_{N \geq 0} \sup_{n,m \geq N} |g_n(\omega) - g_m(\omega)| > 2\varepsilon\} .$$

If $\omega \in \Omega$, let $T_0(\omega) = \inf\{n \geq 0: |g_n(\omega)| > \varepsilon\}$. If $T_0(\omega) < \infty$, let

$$T_1(\omega) = \inf\{n > T_0(\omega): |g_n(\omega) - g_{T_0(\omega)}(\omega)| > \varepsilon\}, \ldots .$$

Let $K = \ell_H^2$ as before and define the K-valued martingales F and G by

$$D_k(\omega) = (d_k(\omega), 0, 0, \ldots) ,$$
$$E_k(\omega) = (e_k(\omega), 0, 0, \ldots) \quad \text{if} \quad k < T_0(\omega) + 1 ,$$
$$E_k(\omega) = (0, \ldots, 0, e_k(\omega), 0, \ldots) ,$$

where $e_k(\omega)$ is in the j-th position, if $j \geq 1$ and $T_{j-1}(\omega) < k < T_j(\omega) + 1$. Note that

$$|E_k(\omega)|_K = |e_k(\omega)| \leq |d_k(\omega)| \leq |D_k(\omega)|_K$$

and $|F_n|_K = |f_n|$. If $T_j \leq n$, then

$$|G_n|_K^2 \geq |g_{T_0}|^2 + |g_{T_1} - g_{T_0}|^2 + \cdots + |g_{T_j} - g_{T_{j-1}}|^2$$

and, by Theorem 4.1, for all $j \geq 0$,

$$\begin{aligned} P(A_\varepsilon) &\leq \lim_{n \to \infty} P(0 \leq T_0 < \cdots < T_j \leq n) \\ &\leq \limsup_{n \to \infty} P(|G_n|_K \geq \varepsilon(j+1)^{1/2}) \\ &\leq 2\|F\|_1/\varepsilon(j+1)^{1/2} = 2\|f\|_1/\varepsilon(j+1)^{1/2} . \end{aligned}$$

Therefore, $P(A_\varepsilon) = 0$ for all $\varepsilon > 0$ and the theorem is proved.

The above proof contains the proof of the following theorem.

THEOREM 5.2. *If f and g are H-valued martingales with respect to $\mathcal{F}$ and g is differentially subordinate to f, then for stopping times satisfying $0 \leq T_0 \leq T_1 \leq \cdots \leq \infty$ and $j, n \geq 0$,*

$$P(T_j \leq n \text{ and } |g_{T_0}|^2 + \sum_{i=1}^{j} |g_{T_i} - g_{T_{i-1}}|^2 \geq 1) \leq 2\|f_n\|_1 .$$

If $\|f\|_1$ is finite, then, by Theorem 5.1, the limit g_∞ of g exists almost everywhere and the g_{T_i} are well defined. In this case, let

$$S(g,T) = [|g_{T_0}|^2 + \sum_{j=1}^{\infty} |g_{T_j} - g_{T_{j-1}}|^2]^{1/2}$$

where T denotes the sequence $(T_j)_{j \geq 0}$. Then, under the conditions of Theorem 5.2,

(5.1) $$P(S(g,T) \geq 1) \leq 2\|f\|_1 ,$$

and, by Theorem 4.2,

(5.2) $$\|S(g,T)\|_p \leq (p^* - 1)\|f\|_p .$$

By the example following the statement of Theorem 4.1 and by Theorem 4.2, the constants in (5.1) and (5.2) are best possible.

REMARK 5.1. The statement of Theorem 5.1 remains true if the condition of differential subordination is replaced by the weaker condition $S(g) \leq S(f)$: Theorem 5.1 in its present form yields the Hilbert-space analogue of Theorem 3 of [10]. This analogue, by the method of Davis [22], then gives the almost everywhere convergence of g under the weaker condition.

REMARK 5.2. The special case $g = f$ of the inequality in Theorem 5.2 can also be deduced from the inequality $P(S_n(f) \geq 1) \leq 2\|f_n\|_1$ applied to the martingale $(f_{T_j \wedge n})_{j \geq 0}$.

If B is a Banach space that is not isomorphic to a Hilbert space, then the conclusion of Theorem 5.1 is no longer valid.

THEOREM 5.3. *Let $1 < p < \infty$. For a Banach space B, the following statements, each to hold for all probability spaces and all f and g as in Theorem 5.1, are equivalent*:

(5.3) $$\|f\|_1 < \infty \Rightarrow g \text{ converges a.e.} ,$$

(5.4) $$P(g^* \geq 1) \leq c(B)\|f\|_1 ,$$

(5.5) $$\|g\|_p \leq c_p(B)\|f\|_p ,$$

(5.6) B *is isomorphic to a Hilbert space* .

See [12] for analogous theorems for B-valued martingales and their transforms, where B belongs to the much larger class of UMD spaces.

We have already proved that (5.6) implies (5.3). The implications (5.3) => (5.4) => (5.5) are proved in the same way as the analogous steps in the proof of Theorem 1.1 of [12]. For the proof of the implication (5.5) => (5.6), see Section 5 of [12], where Kwapień's theorem [34] is used.

6. THREE CLASSES OF BANACH SPACES

Suppose that B is a real or complex Banach space with norm $|\cdot|$ and $1 < p < \infty$. Recall that B is UMD (all B-valued martingale difference sequences are unconditional in $L_B^p([0,1])$) if and only if there is a biconvex function $\zeta: B \times B \to \mathbb{R}$ such that $\zeta(0,0) > 0$ and

$$(6.1) \qquad \zeta(x,y) \le |x + y| \quad \text{if} \quad |x| = |y| = 1 .$$

See [12] and, for a different proof, [17]. The Banach space B is <u>ζ-convex</u> if there is such a function ζ, a condition that does not depend on the choice of p. By replacing "$\zeta(0,0) > 0$" with "$\zeta(0,0) = 1$" in the above statement, we obtain a characterization of Hilbert space [13]. Here is a characterization of spaces isomorphic to a Hilbert space.

THEOREM 6.1. <u>A Banach space B is isomorphic to a Hilbert space if and only if there is a function $\zeta: B \times B \to \mathbb{R}$ such that (6.1) holds, $\zeta(0,0) > 0$, and, for all $x,y,a,b \in B$ with $|a - b| \le |a + b|$, the mapping $\alpha \to \zeta(x + \alpha a, y + \alpha b)$ is convex on $\mathbb{R}$.</u>

A proof resting on Theorem 5.3 can be constructed along the lines of the proof of the equivalence of the UMD condition and the ζ-convexity condition.

To end the paper, we return to the analogy mentioned at its beginning. More than an analogy, the connections are deep and carry over to the B-valued setting where the UMD spaces are the good spaces. There are many consequences for singular integral operators, multiplier theorems, Hardy spaces, and the like. For an introduction, see [17] and [42], two recent and largely nonoverlapping surveys, which contain a number of new results and a large number of references to the literature. Some papers in this setting that have been written or published since 1985 are a part of the list below.

REFERENCES

[1] A. Baernstein, Some sharp inequalities for conjugate functions, Indiana Univ. Math. J. 27 (1978), 833-852.

[2] E. Berkson, T. A. Gillespie, and P. S. Muhly, Abstract spectral decompositions guaranteed by the Hilbert transform, Proc. London Math. Soc. 53 (1986), 489-517.

[3] E. Berkson, T. A. Gillespie, and P. S. Muhly, A generalization of Macaev's theorem to non-commutative L^p-spaces, Integral Equations and Oper. Theory 10 (1987), 164-186.

[4] E. Berkson, T. A. Gillespie, and P. S. Muhly, Analyticity and spectral decompositions of L^p for compact abelian groups, Pacific J. Math. 127 (1987), 247-260.

[5] O. Blasco, Hardy spaces of vector valued functions: Duality, to appear.

[6] O. Blasco, Boundary values of functions in vector-valued Hardy spaces and geometry on Banach spaces, to appear.

[7] J. Bourgain, Some remarks on Banach spaces in which martingale difference sequences are unconditional, Ark. Mat. 21 (1983), 163-168.

[8] J. Bourgain, Extension of a result of Benedek, Calderón and Panzone, Ark. Mat. 22 (1984), 91-95.

[9] J. Bourgain, Vector valued singular integrals and the H^1 - BMO duality, Probability Theory and Harmonic Analysis, J. A. Chao and W. A. Woyczynski, editors, Marcel Dekker, New York (1986), 1-19.

[10] D. L. Burkholder, Martingale transforms, Ann. Math. Stat. 37 (1966), 1494-1504.

[11] D. L. Burkholder, A sharp inequality for martingale transforms, Ann. Probab. 7 (1979), 858-863.

[12] D. L. Burkholder, A geometrical characterization of Banach spaces in which martingale difference sequences are unconditional, Ann. Probab. 9 (1981), 997-1011.

[13] D. L. Burkholder, Martingale transforms and the geometry of Banach spaces, Proceedings of the Third International Conference on Probability in Banach Spaces, Tufts University, 1980, Lecture Notes in Mathematics 860 (1981), 35-50.

[14] D. L. Burkholder, A nonlinear partial differential equation and the unconditional constant of the Haar system in L^p, Bull. Amer. Math. Soc. 7 (1982), 591-595.

[15] D. L. Burkholder, A geometric condition that implies the existence of certain singular integrals of Banach-space-valued functions, Conference on Harmonic Analysis in Honor of Antoni Zygmund, University of Chicago, 1981, Wadsworth International Group, Belmont, California, 1 (1983), 270-286.

[16] D. L. Burkholder, Boundary value problems and sharp inequalities for martingale transforms, Ann. Probab. 12 (1984), 647-702.

[17] D. L. Burkholder, Martingales and Fourier analysis in Banach spaces, C.I.M.E. Lectures, Varenna, Italy, 1985, Lecture Notes in Mathematics 1206 (1986), 61-108.

[18] D. L. Burkholder, A proof of Pelczyński's conjecture for the Haar system, Studia Math., to appear.

[19] D. L. Burkholder, Sharp inequalities for martingales and stochastic integrals, Colloque Paul Lévy, Astérisque, to appear.

[20] F. Cobos, Some spaces in which martingale difference sequences are unconditional, Bull. Polish Acad. of Sci. Math. 34 (1986), 695-703.

[21] T. Coulhon and D. Lamberton, Régularité L^p pour les équations d'évolution, to appear.

[22] B. Davis, Comparison tests for the convergence of martingales, Ann. Math. Stat. 39 (1968), 2141-2144.

[23] B. Davis, On the weak type (1,1) inequality for conjugate functions, Proc. Amer. Math. Soc. 44 (1974), 307-311.

[24] J. L. Doob, Stochastic Processes, Wiley, New York, 1953.

[25] J. L. Doob, Semimartingales and subharmonic functions, Trans. Amer. Math. Soc. 77 (1954), 86-121.

[26] J. L. Doob, Remarks on the boundary limits of harmonic functions, J. SIAM Numer. Anal. 3 (1966), 229-235.

[27] M. Essén, A superharmonic proof of the M. Riesz conjugate function theorem, Ark. Math. 22 (1984), 241-249.

[28] D. L. Fernandez and J. B. Garcia, Interpolation of Orlicz-valued function spaces and U.M.D. property, to appear.

[29] T. W. Gamelin, Uniform Algebras and Jensen Measures, Cambridge University Press, London, 1978.

[30] J. Garcia-Cuerva and J. L. Rubio de Francia, Weighted norm inequalities and related topics, North Holland, Amsterdam, 1985.

[31] D. J. H. Garling, Brownian motion and UMD-spaces, Conference on Probability and Banach Spaces, Zaragoza, 1985, Lecture Notes in Mathematics 1221 (1986), 36-49.

[32] D. J. H. Garling, Random martingale transform inequalities, to appear.

[33] A. N. Kolmogorov, Sur les fonctions harmoniques conjuguées et les séries de Fourier, Fund. Math. 7 (1925), 24-29.

[34] S. Kwapień, Isomorphic characterizations of inner product spaces by orthogonal series with vector valued coefficients, Studia Math. 44 (1972), 583-595.

[35] J. Marcinkiewicz, Quelques théorèmes sur les séries orthogonales, Ann. Soc. Polon. Math. 16 (1937), 84-96.

[36] T. R. McConnell, On Fourier multiplier transformations of Banach-valued functions, Trans. Amer. Math. Soc. 285 (1984), 739-757.

[37] T. R. McConnell, Decoupling and stochastic integration in UMD Banach spaces, Probab. Math. Stat., to appear.

[38] R. E. A. C. Paley, A remarkable series of orthogonal functions I. Proc. London Math. Soc. 34 (1932), 241-264.

[39] A. Pełczyński, Norms of classical operators in function spaces, Colloque Laurent Schwartz, Astérisque 131 (1985), 137-162.

[40] S. K. Pichorides, On the best values of the constants in the theorems of M. Riesz, Zygmund and Kolmogorov, Studia Math. 44 (1972), 165-179.

[41] M. Riesz, Sur les fonctions conjuguées, Math. Z. 27 (1927), 218-244.

[42] J. L. Rubio de Francia, Martingale and integral transforms of Banach space valued functions, Conference on Probability and Banach Spaces, Zaragoza, 1985, Lecture Notes in Mathematics 1221 (1986), 195-222.

[43] J. L. Rubio de Francia, Linear operators in Banach lattices and weighted L^2 inequalities, Math. Nachr. 133 (1987), 197-209.

[44] J. L. Rubio de Francia, F. J. Ruiz, and J. L. Torrea, Calderón-Zygmund theory for operator-valued kernels, Advances in Math. 62 (1986), 7-48.

[45] J. L. Rubio de Francia and J. L. Torrea, Some Banach techniques in vector valued Fourier analysis, Colloq. Math., to appear.

[46] E. M. Stein, Singular Integrals and Differentiability Properties of Functions, Princeton University Press, Princeton, 1970.

[47] E. M. Stein and G. Weiss, On the theory of harmonic functions of several variables I. The theory of H^p-spaces, Acta Math. 103 (1960), 25-62.

[48] T. M. Wolniewicz, The Hilbert transform in weighted spaces of integrable vector-valued functions, Colloq. Math. 53 (1987), 103-108.

[49] F. Zimmermann, On vector-valued Fourier multiplier theorems, to appear.

[50] A. Zygmund, Trigonometric Series II, Cambridge University Press, Cambridge, 1959.

[51] T. Figiel, On equivalence of some bases to the Haar system in spaces of vector-valued functions, Bull. Polon. Acad. Sci., to appear.

SQUARE FUNCTIONS, CAUCHY INTEGRALS, ANALYTIC CAPACITY, AND HARMONIC MEASURE

Peter W. Jones
Yale University
New Haven, CT 06520

Section 0. Introduction

In these notes we will present several examples which link certain geometric properties to square functions and L^p estimates. The first example we consider is the Cauchy integral on curves. It is by now well understood that three objects (at least) play a role in bounding Cauchy integrals: maximal functions, Carleson measures, and square functions. In Section 1 we present a new proof of the L^2 boundedness of the Cauchy integral when the curve is Lipschitz. The novelty here is that we reduce our problem rapidly to a Littlewood-Paley type property of the curve. To be more specific, let Γ be a Lipschitz curve and call $I \subset \Gamma$ dyadic if its projection on $\mathbb{R}$ is dyadic. Let $\beta_0(I)$ equal length $(I)^{-1}$ times the maximal distance of I to a best approximating line. (So if I is a line segment, $\beta_0(I) = 0$, while if I is a "tent", $\beta_0(I) \approx 1$.) then we obtain the estimate (the "Geometric Lemma")

$$\sum_{I \subset J} \beta_0^2(I)|I| \leq C|J|.$$

Roughly speaking, Γ is essentially a line at most scales. Since the Cauchy integral is trivial for lines, our Carleson measure estimate above can be used to bound any "defect" from the linear case. The Carleson measure estimate is in turn obtained by looking at square functions for A', where $\Gamma = \{x + iA(x) : x \in \mathbb{R}\}$.

In Section 2 we give an alternative approach to Guy David's proof of the boundedness of Cauchy integrals on Ahlfors-David (AD) curves [D1]. Recall that γ satisfies the AD condition if

$$m_1(\gamma \cap \{|z-z_0| \le r\}) \le Cr$$

for all $z_0 \in \mathbb{C}$ and $r > 0$. Here m_1 is one dimensional Hausdorff measure. By using a (square function!) theorem due to Garnett, we first show we may consider only the case where each component of $\overline{\mathbb{C}}\setminus\gamma$ is a Lipschitz domain. We then use the known (and "relatively" easy) square function result of Kenig:

$$(1) \qquad \int_{\partial\Omega} |f(z)|^2 ds \le C \iint_{\Omega} |f'(z)|^2 d(z)dxdy.$$

Here f is a holomorphic function on a Lipschitz domain which vanishes at some central point of Ω. By $d(z)$ we denote the distance to $\partial\Omega$. (This last result had previously been used [Jo,Se] to give a very short proof when γ is actually Lipschitz.) By duality we reduce most of the proof to showing that the operator

$$TF(\varsigma) = \iint_{\mathbb{C}\setminus\gamma} \frac{F(z)d(z)}{(z-\varsigma)^2} dxdy$$

is a bounded mapping from $\mathcal{H}$ to $L^2(\gamma)$, where $\mathcal{H}$ is the Hilbert space of complex valued measurable functions satisfying

$$\|F\|_{\mathcal{H}} = \Big(\iint_{\mathbb{C}\setminus\gamma} |F(z)|^2 d(z)dxdy\Big)^{1/2} < \infty.$$

This is accomplished by using (1), the Hardy-Littlewood maximal function, and a strange ad-hoc argument with some auxilliary Carleson measures. As far as I know, this is the only alternative proof of David's theorem we presently have.

It should be pointed out that there is a great overlap between Sections 1 and 2. For example, Lemma 1 of Section 1 is a statement about $\mathcal{H}$ and can easily be seen to be equivalent to Lemma 7 of Section 2. We wish to present two points of view for handling the "square function" space $\mathcal{H}$.

In Section 3 we make some short comments on analytic capacity, a subject which has been revived by Takafumi Murai. If $K \subset \mathbb{C}$ is compact we define $H_0^\infty(K)^c)$ to be the class of analytic functions in $\overline{\mathbb{C}}\backslash K$ which vanish at ∞. Then the analytic capacity of K, $\gamma(K)$, is defined by

$$\gamma(K) = \sup\{|f'(\infty)| : \|f\|_{H_0^\infty(K^c)} \leq 1\}.$$

Here $f'(\infty) = a$ if $f(z) = az^{-1} + O(z^{-2})$ at ∞. We first present some material from Murai's new book [ ], relating analytic capacity to weak type estimates for the Cauchy integral. We then give a short proof of Garnett's theorem: the analytic capacity of the "middle 2/4" Cantor set is zero. This Cantor set plays a central role in obtaining sharp estimates for bounds on the Cauchy integral. Our proof of Garnett's theorem is, of course, via square function estimates. Section 3 ends with a brief discussion of Murai's example of a set with positive analytic capacity but zero Crofton_α length. (There are actually square function estimates lurking in the background of his proof, but we do not discuss these as they are fully covered in Murai's book.)

In Section 4 we turn to harmonic measure on simply connected domains. We first recall a theorem due to Makarov [MaK]. For an increasing function $h(t)$ with $h(0) = 0$ we define as usual the Hausdorff measure Λ_h. For $h(t) = t^\alpha$ we abbreviate $\Lambda_\alpha = m_\alpha$ to be α-dimensional Hausdorff measure. For $c > 0$ let

$$h_c(t) = t \exp\{-c\sqrt{\log \tfrac{1}{t} \log \log \log \tfrac{1}{t}}\}.$$

Theorem (Makarov). <u>If</u> Ω <u>is simply connected and</u> ω <u>is harmonic measure on</u> Ω, <u>then</u>

(a) <u>there is</u> c_0 <u>such that</u>

$$\Lambda_{h_{c_0}}(E) = 0 \Rightarrow \omega(E) = 0.$$

(b) <u>there is</u> $c_1 < c_0$ <u>and an</u> Ω <u>and</u> E <u>such that</u>

$$\Lambda_{h_{c_1}}(E) = 0 \ \ \underline{\text{but}} \ \ \omega(E) = 1.$$

(c) <u>If</u> $\alpha > 1$ <u>there is</u> E <u>such that</u> $\omega(E) = 1$ <u>and</u> $\Lambda_\alpha(E) = 0$.

We call a domain Ω which is a counter-example, in the sense of (b), a <u>Makarov-domain</u>. We present a theorem which says that if $\partial\Omega$ is wiggly enough on all scales, Ω is a Makarov domain. The proof involves square function estimates for $\log f'$, where f is a conformal mapping from the unit disk to Ω.

In Section 5 we present some open problems.

I am greatly indebted to Stephen Semmes for numerous discussions concerning Sections 1 and 2; the results of Section 2 grew out of our joint paper [Jo,Se] and should be viewed as a minor variant of the argument in [Jo,Se].

The loss of José Luis Rubio de Francia came as a great blow. He was not only an excellent mathematician, but an extraordinary person as well. His great warmth and extreme modesty were a model to me. I feel he touched our lives in a very special way.

<u>Section 1. Lipschitz curves</u>

Let $\Gamma = \{x + iA(x) : x \in \mathbb{R}\}$, $A' \in L^\infty$, be a Lipschitz graph in the plane.

The purpose of this section is to give a new proof of the L^2 boundedness of the Cauchy integral on Γ. This was first proved by A.P. Calderón [C] when $\|A'\|_\infty$ is small and later by Coifman, McIntosh, and Meyer [Co,Mc,Me] for the general case; there are now several proofs and generalizations (see e.g. [D1],[D,J],[D,J,Se],[Jo,Se],[Mu]). While the ideas used in this section are almost direct descendants of those used in the previously cited articles, the shortness and elementary nature of our proof may have some attraction. The main idea is to use the geometrical estimates of Lemmata 2 and 4; this seems to be new.

We now fix some notation. Let $\{I_j\}$ denote the collection of all dyadic intervals on $\mathbb{R}$; we denote by I, J, K generic dyadic intervals. The center of I or I_j is written as x_I or x_j. The length of an interval is $|I_j| = d_j$, and we denote by λI the interval with the same center as I and length $|\lambda I| = \lambda|I|$. The domains $\Omega_\pm = \{x + i(A(x) \pm t): x \in \mathbb{R}, t > 0\}$ represent the upper and lower sides of Γ. We define

$$Cf(z) = \lim_{\delta\to 0} \int_{+\Gamma} \frac{f(\zeta)d\zeta}{z+i\delta-\zeta}, \quad z \in \Gamma,$$

to be the Cauchy integral of f. Then Cf admits a holomorphic extension to Ω_+ and on Γ, Cf differs from the usual principal value definition by a multiple of f.

The next lemma is equivalent to the boundedness of C on $L^2(\Gamma)$. The lemma should be compared with the results of Kenig, [K].

Lemma 1. <u>Let</u> $\tilde{z}_j = \tilde{x}_j + i(A(\tilde{x}_j) \pm \tilde{y}_j)$ <u>where</u> $\tilde{x}_j \in I_j$ <u>and</u> $\frac{1}{2}d_j \le \tilde{y}_j < d_j$. <u>Then for any sequence</u> $\{\gamma_j\}$ <u>of complex numbers</u>

$$\Big\|\sum \gamma_j \Big(\frac{d_j}{z-\tilde{z}_j}\Big)^2\Big\|_{L^2(\Gamma)} \le C\Big(\sum|\gamma_j|^2 d_j\Big)^{1/2},$$

where C depends only on $\|A'\|_\infty$.

The following argument shows that Lemma 1 implies the boundedness of the Cauchy integral. This argument is due to Stephen Semmes and greatly shortens a previous proof, due to the author, based on Lemma 1 and the other methods used later in this paper. Let $g \in L^2(\Gamma)$, $\|g\|_{L^2(\Gamma)} \le 1$, and suppose $\sum|\gamma_j|^2 d_j = 1$. Then by Lemma 1, if $\operatorname{Im}(\tilde{z}_j) = A(\tilde{x}_j) + \tilde{y}_j$,

$$C \ge \left|\int_\Gamma g(z)\sum\gamma_j\left(\frac{d_j}{z-\tilde{z}_j}\right)^2 dz\right|$$

$$= 2\pi\left|\sum\gamma_j d_j^2 C'g(\tilde{z}_j)\right|.$$

Taking the supremum over all such $\{\gamma_j\}$ we obtain (see again [K] for similar estimates)

$$(1) \qquad \left(\sum|d_j C'(\tilde{z}_j)|d_j\right)^{1/2} \le C.$$

For a function F vanishing at ∞ we have

$$F(z) = -\int_0^\infty \frac{\partial}{\partial t} F(z+it)dt = \int_0^\infty \frac{\partial^2}{\partial t^2} F(z+it)t\,dt.$$

Applying this formula and Cauchy's theorem we see that for $z \in \Gamma$,

$$2\pi i Cg(z) = 2\pi i\int_0^\infty C''g(z+it)\,t\,dt$$

$$= \int_0^\infty\int_\Gamma \frac{C'g(\zeta+i\,t/2)}{(z+it-(\zeta+i\,t/2))^2}\,d\zeta\,t\,dt$$

$$= \int_{t=0}^\infty\int_{x=-\infty}^\infty \frac{C'g(x+i(A(x)+t/2))(1+iA'(x))}{(z+i\,t/2-(x+iA(x)))^2}\,dxdt$$

$$= \iint_{\Omega_+} \frac{C'g(\varsigma)}{(z-\bar{\varsigma})^2} W(\varsigma)dA(\varsigma).$$

In the above formulae we have set $\bar{\varsigma} = z - i\ t/2$ whenever $\varsigma = z + it$ and $z \in \Gamma$. For the function $W(\varsigma)$ we have the estimate

$$\|W(\varsigma)\|_{L^\infty(Q_j)} \sim d_j$$

where $Q_j = \{x + iy,\ x \in I_j,\ \frac{1}{2}d_j \le y - A(x) < d_j\}$. Writing

$$Cg(z) = \frac{1}{2\pi i} \sum_j \iint_{Q_j} \frac{C'g(\varsigma)}{(z-\bar{\varsigma})^2} W(\varsigma)dA(\varsigma)$$

we see that Cg is a continuous convex combination of functions of the form

$$F = \sum w_j d_j C'g(\tilde{z}_j)\left(\frac{d_j}{z-\tilde{z}_j}\right)^2,$$

where $\|\{w_j\}\|_{\ell^\infty} \le C$ and $\operatorname{Im}(\tilde{z}_j) = A(\tilde{x}_j) - \tilde{y}_j$. But by (1) and Lemma 1, $\|F\|_{L^2(\Gamma)} \le C$, so Minkowski's integral inequality yields $\|Cg\|_{L^2(\Gamma)} \le C$.

It therefore only remains to prove Lemma 1. We accomplish this by using the observation that a Lipschitz graph may be well approximated locally by straight lines. This "goodness of fit" is given quantitatively in Lemma 4 by a quadratic Carleson measure estimate. We first take the Calderón decomposition of $a(x) = A'(x)$. Calderón's decomposition yields the representation

$$a(x) = \sum_I a_I \psi_I(x)$$

where the sum is taken over all dyadic I, where ψ_I is supported on $3I$, $|\psi'_I(x)| \le |I|^{-1}$, and where $\int \psi_I dx = 0$. A well known argument using Plancherel's formula yields the BMO type estimate

$$\sup_J \left(\frac{1}{|J|} \sum_{I \subseteq J} |a_I|^2 |I|\right)^{1/2} \leq C\|a\|_\infty. \tag{2}$$

See e.g. [U]. For each dyadic I let

$$\alpha_I = \sum_{|J| \geq |I|} a_J \psi_J(x_I)$$

denote the "average slope" of Γ near I, and define

$$\beta_0(I) = \sup_{x \in 3I} \frac{1}{|I|} |A(x) - A(x_I) - \alpha_I(x-x_I)|$$

the main idea of this section is the observation that $\beta_0(I)$ is, on average, small. More specifically, we have the following Carleson condition.

Lemma 2 (The Geometric Lemma). $\sup_J \left(\frac{1}{|J|} \sum_{I \subseteq J} \beta_0(I)\right)^{1/2} \leq C\|a\|_\infty.$

Proof. For a dyadic interval I we define I^n to be that dyadic interval such that $I \subset I^n$, $|I^n| = 2^n|I|$, $n \geq 0$. For each $n \geq 0$ we also let $\tilde{I}^n$ be that dyadic interval which maximizes $|a_J|$ under the constraint $|J| = 2^n|I|$, $3J \cap I \neq \phi$. For $n > 0$ let $I_n \subset 3I$ be that dyadic interval which maximizes $|a_J|$ under the constraint $|J| = 2^{-n}|I|$. We will demonstrate that

$$\beta_0(I) \leq C \sum_{n=0}^{\infty} |a_{\tilde{I}^n}| \cdot 2^{-n} \tag{3}$$

$$+ C \sum_{n=0}^{\infty} |a_{I_n}| 2^{-n}$$

To this end, we may suppose without loss of generality that $A(x_I) = 0$. Fix $x \in 3I$ and write $A(x) = \int_{x_I}^{x} a(t)dt$, so that

$$A(x) - \alpha_I(x-x_I) = \int_{x_I}^{x} (a(t) - \alpha_I)dt$$

$$= \int_{x_I}^{x} \sum_{|J|\geq I} a_J(\psi_J(t) - \psi_J(x_I))dt$$

$$+ \int_{x_I}^{x} \sum_{|J|<|I|} a_J\psi_J(t)dt$$

$$= E_1 + E_2$$

Since only three intervals J of fixed length satisfy $I \cap 3J \neq \phi$, the estimate $|\psi_J'| \leq |J|^{-1}$ is applied to obtain

$$|E_1| \leq \int_{x_I}^{x} \sum_{n=0}^{\infty} \sum_{|J|=2^n|I|} |a_J||\psi_J(t) - \psi_J(x_I)|dt$$

$$\leq 3\int_{x_I}^{x} \sum_{n=0}^{\infty} |a_{\tilde{I}^n}|\cdot 2^{-n}dt$$

$$\leq 3 \sum_{n=0}^{\infty} |a_{\tilde{I}^n}|\cdot 2^{-n}|I|.$$

On the other hand, for each fixed $n > 0$ at most 6 intervals J with $|J| = 2^{-n}|I|$ satisfy $\int_{x_I}^{x} \psi_J(t)dt \neq 0$, and consequently,

$$|E_2| = \left|\int_{x_I}^{x} \sum_{n=0}^{\infty} \sum_{|J|=2^{-n}|I|} a_J\psi_J(t)dt\right|$$

$$\le 6 \sum_{n=1}^{\infty} |a_{I_n}| \cdot \tfrac{3}{2} 2^{-n} |I|.$$

The last inequality follows from the observation that $\left|\int_{x_I}^{x} \psi_J(t)dt\right| \le \frac{3}{2}|J|$.

Combining the above estimates for E_1 and E_2 yields estimate (3).

To prove Lemma 2 it is now sufficient to prove the estimates

$$(4) \qquad \left(\frac{1}{|J|} \sum_{I \subseteq J} |a_{\tilde{I}^n}|^2 |I|\right)^{1/2} \le c 2^{n/2} \|a\|_\infty, \quad n \ge 0$$

and

$$(5) \qquad \left(\frac{1}{|J|} \sum_{I \subseteq J} |a_{I_n}|^2 |I|\right)^{1/2} \le c 2^{n/2} \|a\|_\infty, \quad n \ge 0.$$

For then the weights 2^{-n} in estimate (3) allow us to use Minkowski's inequality to estimate $\left(\frac{1}{|J|} \sum_{I \subseteq J} \beta_0(I)^2 |I|\right)^{1/2}$ by the geometric sum $\sum_{n=0}^{\infty} c 2^{-n/2} \|a\|_\infty$. Fix n and notice that since $|\tilde{I}^n| = 2^n |I|$, $\tilde{I}^n \subset 3I^n$, each coefficient $a_{\tilde{I}^n}$ is counted <u>at most</u> $3 \cdot 2^n$ <u>times</u> in the sum in (4).

Consequently

$$(4) = \left(\frac{1}{|J|} \sum_{I \subseteq J} |a_{\tilde{I}^n}|^2 \cdot 2^{-n} |\tilde{I}^n|\right)^{1/2}$$

$$\le \left(\frac{1}{|J|} \sum_{I \subset 3J^n} |a_K|^2 \cdot |K|\right)^{1/2}$$

$$\le 2^{n/2} \left(\frac{3}{|J^n|} \sum_{I \subset 3J^n} |a_K|^2 \cdot |K|\right)^{1/2}$$

$$= c2^{n/2}\|a\|_\infty,$$

the last line following from (2). Similarly, we see in estimating (5) that each term a_{I_n} is counted at most three times in the sum, so

$$(5) \;=\; \left(\frac{1}{|J|}\sum_{I\subseteq J} |a_{I^n}|^2 2^n |I_n|\right)^{1/2}$$

$$\le 2^{n/2}\left(\frac{1}{|J|}\sum_{K\subseteq 3J} 3\cdot|a_K|^2|K|\right)^{1/2}$$

$$= c2^{n/2}\|a\|_\infty.$$

The proof of Lemma 2 is complete.

We now define for $n \ge 1$

$$\beta_n(I) = \sup_{\frac{|x-x_I|}{|I|}\sim 2^n} \frac{1}{2^n|I|}\,|A(x)-A(x_I) - \alpha_I(x-x_I)|.$$

Lemma 3. $\sup_J\left(\frac{1}{|J|}\sum_{I\subseteq J}\beta_n(I)^2|I|\right)^{1/2} \le c2^{n/2}\|a\|_\infty.$

Proof. By the triangle inequality,

$$\beta_n(I) \le 2\beta_0(I^n) + |\alpha_{I^n} - \alpha_I|.$$

Now each interval J can equal I^n for exactly 2^n intervals I, so

$$\left(\frac{1}{|J|}\sum_{I\subseteq J}\beta_0(I^n)^2|I|\right)^{1/2} = \left(\frac{2^n}{|J^n|}\sum_{I\subseteq J}\beta_0(I^n)^2\cdot 2^{-n}|I^n|\right)^{1/2}$$

$$\le \left(\frac{1}{|J^n|}\sum_{K\subseteq J^n} 2^n\beta_0(K)^2|K|\right)^{1/2}$$

$$\le c2^{n/2}\|a\|_\infty$$

by Lemma 2. To handle the other piece notice that by definition,

$$|\alpha_{I^n} - \alpha_I| \le 3\sum_{k=0}^{n-1} |a_{\tilde{I}^k}| + \sum_{|J|\ge|I^n|} |a_J||\psi_J(x_{I^n}) - \psi_J(x_I)|$$

$$\le 3\sum_{k=0}^{n-1} |a_{\tilde{I}^k}| + 3\sum_{k=n}^{\infty} |a_{\tilde{I}^k}|\cdot 2^{n-k}$$

Now by (4) we see that

$$\left(\frac{1}{|J|}\sum_{I\subset J} |\alpha_{I^n} - \alpha_I|^2 |I|\right)^{1/2}$$

$$\le c\|a\|_\infty \left\{\sum_{k=0}^{n-1} 2^{k/2} + \sum_{k=n}^{\infty} 2^{n-k}\cdot 2^{k/2}\right\}$$

$$\le c2^{n/2}\|a\|_\infty,$$

and this proves Lemma 3.

Define $\beta(I) = \sum_{n=0}^{\infty} \beta_n(I)2^{-3n/4}$. The next lemma follows directly from the definition of $\beta_n(I)$ and from Lemmata 2 and 3.

Lemma 4. <u>The</u> $\beta(I)$ <u>satisfy</u>

$$(6) \qquad |A(x)-A(x_I) - \alpha_I(x-x_I)| \le \beta(I)|I|\left(1 + \frac{|x-x_I|}{|I|}\right)^{7/4}$$

<u>and</u>

$$(7) \qquad \sup_J \left(\frac{1}{|J|}\sum_{I\subset J} \beta(I)^2|I|\right)^{1/2} \le c\|a\|_\infty .$$

We now use Lemma 4 to prove Lemma 1. Set $F_j(z) = \left(\frac{d_j}{z-\tilde{z}_j}\right)^2$ and write

$(z_j = \tilde{x}_j + iA(\tilde{x}_j))$

$$F_j(x+iA(x)) = F_j(z_j + (x-\tilde{x}_j)(1 + i\alpha_{I_j}))$$

$$+ (F_j(x+iA(x)) - F_j(z_j + (x-\tilde{x}_j)(1 + i\alpha_{I_j})))$$

$$= G_j(x) + E_j(x),$$

so that G_j is the restriction to a line of F_j, and E_j is an error term. By Cauchy's theorem, $\int_{-\infty}^{\infty} G_j(x)dx = 0$, and by the form of F_j, we have the estimates

$$|G_j(x)| \le C\left(1 + \frac{|x-x_j|}{d_j}\right)^{-2},$$

$$|G_j'(x)| \le Cd_j^{-1}\left(1 + \frac{|x-x_j|}{d_j}\right)^{-3}.$$

It is an elementary calculation to verify that for any collection of functions G_j satisfying the above three properties,

$$\|\sum\gamma_j G_j\|_{L^2(\mathbb{R})} \le C(\sum|\gamma_j|^2 d_j)^{1/2}.$$

We need therefore only handle the error terms E_j. By (6) and the form of F_j,

$$|E_j(x)| \le c\beta(I_j)\left(1 + \frac{|x-x_j|}{d_j}\right)^{-5/4}.$$

To see this, merely compute derivatives of F_j, and integrate in the crudest fashion. Now set $\varphi(x) = (1 + |x|)^{-5/4}$ and let $\varphi_t(x) = t^{-1}\varphi(\frac{x}{t})$, $t > 0$, define an approximation to the identity. We also define $g(x + it) = \varphi_t * g(x)$, fix $g \in L^2(\mathbb{R})$, $g \ge 0$, $\|g\|_{L^2(\mathbb{R})} = 1$, and estimate $\|\sum\gamma_j E_j\|_{L^2(\mathbb{R})}$ by duality:

$$\int_{\mathbb{R}} g|\sum\varphi_j E_j|dx \le C\sum|\gamma_j|\beta(I_j)d_j g(x_j + id_j)$$

$$\leq C(\sum|\gamma_j|^2 d_j)^{1/2}(\sum g(x_j + id_j)^2\beta(I_j)^2 d_j)^{1/2}.$$

Let $\|\mu\|_C$ denote the Carleson norm of a Carleson measure μ on $\mathbb{R}^2_+$. Then the last sum on the right above can be estimated by (δ_ζ = Dirac mass at ζ)

$$\sum g(x_j + id_j)^2\beta(I_j)^2 d_j = \iint_{\mathbb{R}^2_+} g(z)^2 \sum\beta(I_j)^2 d_j \delta_{x_j+id_j}(z)$$

$$\leq C\|g\|^2_{L^2(\mathbb{R})} \|\sum\beta(I_j)^2\delta_{x_j+id_j}\|_C$$

$$\leq C\|g\|^2_{L^2(\mathbb{R})} \|a\|_\infty,$$

the last line coming from (7). (For a derivation of the penultimate inequality, see e.g. [G2], page 33.)

The author's student X. Fang has proved an $\mathbb{R}^n$ version of Lemmata 2 and 4 where $\beta_0(I)$ is not defined in an L^∞ matter, but in some L^p fashion. He can then use the argument of this section to give another proof of the boundedness of the Cauchy integral for Lipschitz surfaces.

Section 2. We begin with some "easier" results, the proofs of which can be found in [K] and [Je,K]. (By the remark after Lemma 2, we could use the "easier still" argument in [Jo,Se]. However, the point of this note is to see how the proof of Guy David's theorem devolves naturally from knowledge of square function estimates.) Recall that for a curve γ, $d(z)$ = distance from z to γ. A positive measure μ is a Carleson measure if for all $z_0 \in \gamma$ and all $r \leq \text{diameter}(\gamma)$,

$$\mu(\{|z-z_0| \leq r\}) \leq Ar,$$

for some constant A. If γ is a bounded Jordan curve let z_γ denote a point in the bounded component of $\mathbb{C}\backslash\gamma$ of almost maximal distance to γ.

Lemma 1. <u>Suppose</u> γ <u>is a bounded chord arc curve and suppose</u> f <u>is holomorphic inside</u> γ <u>and</u> f <u>vanishes at</u> z_γ. <u>Then if</u> $\mathcal{D}_+$ = <u>inside of</u> γ,

$$\int_\gamma |f|^2 ds \le c \iint_{\mathcal{D}_+} |f'(z)|^2 d(z) dxdy.$$

<u>If</u> f <u>is holomorphic in</u> $\mathcal{D}_-$ = <u>outside of</u> γ <u>and if</u> $f(\infty) = 0$,

$$\int_\gamma |f|^2 ds \le c \iint_{\mathcal{D}_-} |f'(z)|^2 d(z) dxdy.$$

Lemma 2. <u>With</u> γ <u>and</u> $\mathcal{D}_\pm$ <u>as above</u>,

$$\iint_{\mathcal{D}_\pm} |f(z)|^2 d\mu(z) \le C_0 \|\mu\|_C \int_\gamma |f|^2 ds$$

<u>for any holomorphic</u> f <u>and any</u> μ <u>supported in</u> $\{z: \text{distance}(z,\gamma) \in C$ diameter$(\gamma)\}$.

Here $\|\mu\|_C$ is the Carleson norm of μ. It is an amusing problem to give a "two line" proof of Lemma 1. Such a proof for Lemma 1, in the special case of Lipschitz curves, is given in [Jo,Se]. We also use a weak form of Zinsmeister's theorem from [Z].

Lemma 3. <u>Suppose</u> γ <u>is</u> AD <u>and</u> f <u>is a conformal mapping of</u> $\mathbb{R}^2_+$ <u>onto a component of</u> $\overline{\mathbb{C}}\setminus\gamma$. <u>Then</u> $\log f' \in BMO(\mathbb{R})$ <u>with norm depending only on the</u> AD <u>constant of</u> γ.

Now fix our AD curve γ and let Ω_j be the components of $\overline{\mathbb{C}}\setminus\gamma$. Without loss of generality, all Ω_j are bounded. (This does not really play any role in the proof.) Let f_j denote a conformal mapping of $\mathbb{R}^2_+$ onto Ω_j. By Garnett's version of the corona construction (see e.g. Garnett [G2], page 348) and Lemma 3, we can decompose $\mathbb{R}^2_+$ into disjoint chord-arc domains $\mathcal{O}_{j,k}$ (with chord arc constant bounded by 5) such that

(1) Arc length on $\bigcup_k \partial\mathcal{O}_{j,k}$ is a Carleson measure of norm $C(\epsilon)$

and

(2) $|\log f_j'(z_1) - \log f_j'(z_2)| \le \epsilon$ for all $z_1, z_2 \in \mathcal{O}_{j,k}$.

Taking ϵ small enough (with respect to 5!) we then have by (2) that $f(\mathcal{O}_{j,k}) = \mathcal{D}_{j,k}$ is a chord arc domain with constant less than 6. Conditions (1) and (2) show there are sets $E_{j,k} \subset \gamma \bigcap \{z: \text{distant}(z,\mathcal{D}_{j,k}) \le C\ \text{diameter}(\mathcal{D}_{j,k})\}$ such that $m_1(E_{j,k}) \ge c\ \text{diameter}(\mathcal{D}_{jk})$ and $\|\Sigma\chi_{E_{j,k}}\|_{L^\infty(\gamma)} \le C$. Since a.e. every point in γ can be in the boundary of only two domains Ω_j,

$$\gamma \cup \bigcup_{j,k} \partial\mathcal{D}_{j,k}$$

is also an AD curve. Now this remark can be taken even further. Since each $\mathcal{O}_{j,k}$ is a Carleson corona domain, it is a straightforward (if not short) exercise to prove that each $\mathcal{O}_{j,k}$ can be further decomposed into $\bigcup_\ell \mathcal{O}_{j,k,\ell}$ where each $\mathcal{O}_{j,k,\ell}$ is a Lipschitz domain with Lipschitz constant $\le C_0$, and such that $\bigcup_\ell \partial\mathcal{O}_{j,k,\ell}$ is an AD curve. Taking $\mathcal{D}_{j,k,\ell} = f_j(\mathcal{O}_{j,k,\ell})$ we see that $\mathcal{D}_{j,k,\ell}$ is a Lipschitz domain (if ϵ is small) and

$$\gamma \cup \bigcup_{j,k,\ell} \partial\mathcal{D}_{j,k,\ell}$$

is an AD curve. We state this as a proposition.

Proposition 4. <u>We may assume that the components Ω_j of $\overline{\mathbb{C}}\setminus\gamma$ are bounded Lipschitz domains with Lipschitz constants $\le C_0$.</u>

We now let z_j denote the "center" of Ω_j, i.e. a point for which $d(z_j) \geq 1/2 \sup_{z\in\Omega_j} d(z)$. Setting $d_j = d(z_j)$ we have by Proposition 4,

$$(3) \qquad d_j \sim \text{diameter}(\Omega_j),$$

and the measure $\sum_j d(z_j)\delta_{z_j}$ satisfies the Carleson condition,

$$(4) \qquad \sum_{|z_j - z| \geq r} d(z_j)\delta_{z_j} \leq Ar \quad \text{for all } z \in \gamma.$$

Now let $g \in L^2(\gamma)$ and let Cg be its Cauchy integral on $\overline{\mathbb{C}}\backslash\gamma$. Then we must show that the boundary values of Cg on $\partial\Omega_j$ satisfy

$$\sum_j \int_{\partial\Omega_j} |Cg|^2 ds \leq C \int_\gamma |g|^2 ds.$$

By condition (3) we thus see we must bound the sums

$$(5) \qquad \sum_j \int_{\partial\Omega_j} |Cg - Cg(z_j)|^2 ds$$

and

$$(6) \qquad \sum_j \int_{\partial\Omega_j} |Cg(z_j)|^2 ds.$$

We first attach (5). (This is the main part of the proof.) By Lemma 1 and Proposition 4,

$$(7) \qquad \int_{\partial\Omega_j} |Cg - Cg(z_j)|^2 ds \leq c \iint_{\Omega_j} |Cg'(z)|^2 d(z)\,dxdy.$$

We now follow the lead from [Jo,Se]. Let $\mathcal{H}$ denote the Hilbert space of measurable complex valued functions F on Ω satisfying

$$\|F\|_{\mathcal{H}} = \left(\iint |F(z)|^2 d(z)\,dxdy\right)^{1/2} < \infty,$$

and let $\langle\cdot,\cdot\rangle_{\mathcal{H}}$ denote the inner product on $\mathcal{H}$. Then from (7) we see we must control $\|Cg'\|_{\mathcal{H}}$. This is the main point of the argument. The differences from the short argument of [Jo,Se] are maddening but essentially technical. The one idea presented here is to replace the study of the special points z_j by auxilliary points $z_{j,n}$, $z^*_{j,n}$ and then use Lemma 9 to obtain Corollary 10. If $\mathbb{B}$ denotes the unit ball of $\mathcal{H}$, by duality we obtain $\|Cg'\|_{\mathcal{H}} = \sup_{F\in\mathbb{B}} |\langle Cg',F\rangle_{\mathcal{H}}|$. since $Cg'(z) = \int_\Gamma \frac{g(\zeta)d\zeta}{(\zeta-z)^2}$, we see that for $F \in \mathbb{B}$ with compact support in Ω,

$$\left|\langle Cg',F\rangle_{\mathcal{H}}\right| = \left|\int_\Gamma g(\zeta)\left\{\iint_\Omega \frac{F(z)d(z)dxdy}{(\zeta-z)^2}\right\}d\zeta\right|$$

$$= \left|\int_\Gamma g(\zeta)TF(\zeta)d\zeta\right| .$$

Our goal is thus to see that the mapping

$$TF(\zeta) = \iint_\Omega \frac{F(z)d(z)dxdy}{(\zeta-z)^2}$$

is a bounded mapping from $\mathcal{H}$ to $L^2(\gamma)$. The proof of this is quite simple when γ is Lipschitz (see [Jo,Se]), but there are technical problems for the general AD case. We cope as best we can.

Let $T_jF = T(\chi_{\Omega_j}\cdot F)$ so that $T = T_j + S_j = T_j + (T-T_j)$. We will need

Lemma 5. $\int_{\partial\Omega_j} |T_jF(z_j)|^2 d_j \le C\|\chi_{\Omega_j}F\|^2_{\mathcal{H}}$.

plus the simpler

Lemma 6. $|T_jF(z_j)|^2 d_j \le C\|\chi_{\Omega_j}F\|^2_{\mathcal{H}}$.

Now Lemma 6 is actually false as it stands, but this is easily rectified. Let $\{Q_\ell\}$ denote the Whitney decomposition of Ω. Then by splitting F into two functions, we may assume $F \equiv 0$ on the right half of each Q_ℓ. Since z_j may be picked to be in the right half of a Q_ℓ, we may assume without loss of generality that distance $(z_j$, support $F) \geq cd_j$. Then Lemma 6 follows trivially from (3).

Since S_jF is holomorphic on Ω_j, Lemma 1 yields

$$\int_{\partial\Omega_j} |S_jF|^2 ds \leq C \iint_{\Omega_j} |S_jF'(z)|^2 d(z)\,dxdy$$
$$+ C|S_jF(z_j)|^2 d_j .$$

We therefore require $(\|F\|_{\mathcal{H}} = 1)$

Lemma 7. $\sum_j \iint_{\Omega_j} |S_jF'(z)|^2 d(z)\,dxdy \leq C$.

and

$$\sum_j |S_jF(z_j)|^2 d_j \leq C.$$

By Lemma 6, the last inequality is equivalent to

Lemma 8. $\sum_j |TF(z_j)|^2 d_j \leq C$.

The proof of inequality (5) has thus been reduced to the proofs of Lemmas 5, 7, and 8. The proofs of the first two are straightforward, as in [Jo,Se], bu the proof of Lemma 8 is bothersome.

By invoking Lemma 1, we see that Lemma 5 follows from Lemma 7 (let $\gamma = \partial\Omega_j$). Turning to the proof of Lemma 7 we fix $w \in \Omega_j$ and write

$$|S_jF'(w)| = 2\iint_{\Omega_j^c} \frac{F(z)d(z)}{(z-w)^3}\,dxdy$$
$$\leq 2\iint_{\Omega_j^c} \frac{|F(z)|d(z)^{1/2}}{|\bar{z}-w|^{5/2}}\,dxdy$$

$$\leq 2 \iint_{|z-w|\geq d(w)} \frac{|F(z)|d(z)^{1/2}}{|z-w|^{5/2}} dxdy$$

$$\leq Cd(w)^{-1/2}\mathfrak{M}(Fd^{1/2})(w),$$

where $\mathfrak{M}$ is the classical Hardy-Littlewood operator. The first inequality above comes from the elementary observation that $d(z) \leq |z-w|$, $z \in \Omega_j^c$. Then

$$\sum_j \iint_{\Omega_j} |S_jF'(z)|^2 d(z)dxdy$$

$$\leq C\iint_{\mathbb{C}} d(z)^{-1}\mathfrak{M}(Fd^{1/2})^2(z)d(z)dxdy$$

$$\leq C'\iint_{\mathbb{C}} |F(z)|^2 d(z)dxdy = C'\|F\|_{\mathcal{H}}^2,$$

so Lemma 7 has been verified.

We now turn our attention to Lemma 8. The idea here is to estimate the sum $\sum_j |TF(z_j)|^2 d_j$ by using the triangle inequality M times, where M is a large positive integer. The sum will then feed back upon itself. We define a sequence of points starting at z_j. We first set $z_j = z_{j,0}$. For each j we pick a component Ω_k such that diameter $(\Omega_k) \geq 2$ diameter (Ω_j), $d_k \geq d_j$, and distance $(\Omega_j,\Omega_k) \leq Cd_j$. (Such an Ω_k exists by the AD condition.) Then pick $z_{j,0}^* \in \Omega_k$ to satisfy $d(z_{j,0}^*) = d(z_{j,0}) = d_j$, $|z_{j,0}^* - z_{j,0}| \leq Cd_j$. (Such a point exists by the chord arc condition on Ω_k.) Then define $z_{j,1} = z_k =$ "center Ω_k". We now repeat, setting $z_{j,1}^* = z_{k,0}^* = (z_k)^*$, $z_{j,2} = z_{k,1}$. In this fashion we obtain

$$z_j = z_{j,0} \to z_{j,0}^* \to z_{j,1} \to z_{j,1}^* \to z_{j,2} \to \cdots$$

Then $z_{j,n}^*$ and $z_{j,n}$ lie in the same component of Ω, and

$$(8) \qquad d(z_{j,n+1}) \geq 2d(z_{j,n}) \quad \text{and}$$
$$|z_{j,n} - z^*_{j,n}| \leq cd(z_{j,n}).$$

Each $z_{j,n}$ is the center of some component of Ω. By the triangle inequality,

$$(9) \qquad \Big(\sum_j |TF(z_j)|^2 d_j\Big)^{1/2} \leq \sum_{n=0}^{M-1} \Big(\sum_j |TF(z_{j,n}) - TF(z^*_{j,n})|^2 d_j\Big)^{1/2}$$
$$+ \sum_{n=0}^{M-1} \Big(\sum_j |TF(z^*_{j,n}) - TF(z_{j,n+1}|^2 d_j\Big)^{1/2}$$
$$+ \Big(\sum_j |TF(z_{j,M})|^2 d_j\Big)^{1/2}$$

We now invoke a geometric lemma on the relation between z_j, $z_{j,M}$, and d_j.

Lemma 9. <u>If</u> M <u>is</u> <u>large</u> <u>enough</u>,

$$\sum_{\substack{j \\ z_{j,M}=z_k}} d_j \leq 1/4\, d_k.$$

Proof. Let $\mathcal{F}_n = \{z_j : z_{j,M-n} = z_k\}$. Then by (8) we can sum a geometric series to obtain

$$\mathcal{F}_n \subset \{z : |z - z_k| \leq Cd_k\}.$$

The AD condition for γ then yields

$$\sum_{z_j \in \mathcal{F}_n} d_j \leq Cd_k.$$

Exactly the same argument gives us

$$\sum_{m=1}^{n} \sum_{z_j \in \mathcal{F}_m} d_j \leq Cd_k$$

because the collections $\mathcal{F}_m$ are disjoint.

Repeating this on the collection $\mathcal{F}_m$ we see that

$$\sum_{\ell=1}^{m-1} \sum_{z_j \in \mathcal{F}_\ell} d_j \le C \sum_{z_j \in \mathcal{F}_m} d_j,$$

or, setting $a_n = \sum_{z_j \in \mathcal{F}_n} d_j$,

$$\sum_{n=1}^{m-1} a_n \le Ca_m \quad , \quad 2 \le m \le M. \tag{10}$$

It is an exercise to see that this yields the estimate

$$a_n \le A(1-\delta)^{m-n} a_m,$$

where A and δ depend only on the value of C in (10). Taking M large enough so that $A(1-\delta)^{M-1} < 1/4$, we obtain the proof of Lemma 9.

Corollary 10. If M is large enough,

$$\Big(\sum_j |TF(z_j)|^2 d_j\Big)^{1/2} \le 2 \sum_{n=0}^{M-1} \Big(\sum_j |TF(z_{j,n}) - TF(z^*_{j,n})|^2 d_j\Big)^{1/2}$$

$$+ 2 \sum_{n=0}^{M-1} \Big(\sum_j |TF(z^*_{j,n}) - TF(z_{j,n+1})|^2 d_j\Big)^{1/2}.$$

Corollary 10 follows immediately from (9) and Lemma 9 because

$$\sum_j |TF(z_{j,M})|^2 d_j = \sum_k \sum_{z_{j,M}=z_k} |TF(z_k)|^2 d_j$$

$$\le 1/4 \sum_k |TF(z_k)|^2 d_k.$$

We will also require the elementary

Lemma 11. $\mu = \Sigma d_j \delta_{z^*_{j,o}}$ is a Carleson measure, and $\|\mu\|_C \le A$.

The proof of Lemma 8 has been reduced via Corollary 10 to proving that for each n,

$$\sum_j |TF(z_{j,n}) - TF(z^*_{j,n})|^2 d_j \le C$$

and

$$\sum_j |TF(z^*_{j,n}) - TF(z_{j,n+1})|^2 d_j \le C.$$

Since by the proof of Lemma 9, $\sum_{\substack{j \\ z_{j,n}=z_k}} d_j \le Cd_k$, these two estimates will follow from

(11) $$\sum_j |TF(z_{j,0}) - TF(z^*_{j,0})|^2 d_j \le C$$

and

(12) $$\sum_j |TF(z^*_{j,0}) - TF(z_{j,1})|^2 d_j \le C$$

We attack (11) first. The proof is the same as the proof of Lemma 7. As in the remark after Lemma 6, we may suppose that distance $(z^*_{j,0}$, support $F) \ge cd_j$. The inequality $d(z) \le A|z-z_j|$ clearly holds whenever $|z-z_j| \ge cd_j$. Consequently

$$\begin{aligned} |TF(z_{j,0}) - TF(z^*_{j,0})| &\le A \iint_\Omega \frac{|F(z)|d(z)|z_{j,0}-z^*_{j,0}|}{|z-z_{j,0}|^3}\,dxdy \\ &\le A' \iint_{\Omega\cap\{|z-z_j|>cd_j\}} \frac{|F(z)|d(z)^{1/2}d_j}{|z-z_j|^{5/2}}\,dxdy \\ &\le A'' d_j^{1/2}\mathfrak{M}(Fd(z)^{1/2})(z_j), \end{aligned}$$

where $\mathfrak{M}$ is again the Hardy-Littlewood operator. The inequality persists in a neighborhood of z_j, so that in fact

$$\begin{aligned} &|TF(z_{j,0}) - TF(z^*_{j,0})|^2 d_j \\ &\le C \iint_{|z-z_j|<1/2\, d_j} \mathfrak{M}(Fd(z)^{1/2})^2\,dxdy \end{aligned}$$

Since the disks $\{|z-z_j| < 1/2\, d_j\}$ are disjoint, the sum in (11) is bounded by

$$C \iint_{\mathbb{C}} \mathfrak{M}(Fd^{1/2})^2 dxdy \le C' \iint_{\mathbb{C}} |F|^2 d(z)dxdy$$

$$\le C,$$

so (11) has been handled.

Now to attack (12) we fix Ω_k and let $\mathcal{F}_k = \{z_j : z^*_{j,0} \in \Omega_k\}$. Writing $T = T_k + S_k$ as before, we must bound

(13) $$\sum_k \sum_{z_j \in \mathcal{F}_k} |T_k F(z^*_{j,0}) - T_k F(z_{j,1})|^2 d_j$$

and

$$\sum_k \sum_{z_j \in \mathcal{F}_k} |S_k F(z^*_{j,0}) - S_k F(z_{j,1})|^2 d_j$$

By Lemmas 1, 2, 7, and 11, the second sum is bounded, so we need only bound (13). In a bizarre twist we bounce backward from $z^*_{j,0}$ to $z_{j,0}$! Now since $\sum_{z_j \in \mathcal{F}_k} d_j \le Cd_k$ and since $|T_k F(z_k)|^2 d_k \le c \iint_{\Omega_k} |F(z)|^2 d(z)dxdy$ (Lemma 6), to bound (13) we need only prove that for each k,

(14) $$\sum_{z_j \in \mathcal{F}_k} |T_k F(z_{j,0}) - T_k F(z^*_{j,0})|^2 d_j \le c \iint_{\Omega_k} |F(z)|^2 d(z)dxdy$$

and

(15) $$\sum_{z_j \in \mathcal{F}_k} |T_k F(z_{j,0})|^2 d_j \le c \iint_{\Omega_k} |F(z)|^2 d(z)dxdy$$

But (14) follows immediately from setting $\gamma = \partial\Omega_k$ in (11), and (15) follows from Lemmas 1, 2, and 5 because by (4), $\Sigma d_j \delta_{z_j}$ is a Carleson measure. Therefore inequality (5) holds.

Though we are almost out of energy, the proof of (6) will follow easily from what we have already done. by exactly the reasoning of Corollary 10 plus inequalities (11) and (12) we see that to prove (6) it suffices to show

(16) $$\sum_j |Cg(z_{j,0}) - Cg(z^*_{j,0})|^2 d_j \le C$$

and

(17) $$\sum_j |Cg(z^*_{j,0}) - Cg(z_{j,1})|^2 d_j \le C.$$

Now since $\sum_j d_j \delta_{z^*_{j,0}}$ is a Carleson measure, (17) follows from Lemmas 1, 2, and inequality (7). (We just spent all our time proving that $\|Cg'\|_{\mathcal{H}} \le C$.)

To prove (16) let $E_j \subset \gamma \cap \{|z-z_j| < 2d_j\}$ be sets such that $m_1(E_j) \ge cd_j$ and $\|\sum \chi_{E_j}\|_{L^\infty(\gamma)} \le C$. Then

$$|Cg(z_{j,0}) - Cg(z^*_{j,0})| \le Cd_j \int_j \frac{|g(\zeta)|}{|z_j-\zeta|^2} ds(\zeta)$$

$$\le C \sup_{r>0} \frac{1}{r} \int_{\gamma\cap\{|\zeta-z|\le r\}} |g(\zeta)| ds(\zeta)$$

$$= C\tilde{\mathfrak{m}}g(z) \quad \text{for any} \quad z \in E_j.$$

Now by standard real variable theory $\|\tilde{\mathfrak{m}}g\|_{L^2(\gamma)} \le c\|g\|_{L^2(\gamma)}$, so

$$\sum_j |Cg(z_{j,0}) - Cg(z^*_{j,0})|^2 d_j$$

$$\le C\sum_j \int_{E_j} |\tilde{\mathfrak{m}}g(\zeta)|^2 ds(\zeta)$$

$$= C \int_\gamma |\tilde{\mathfrak{m}}g|^2 \sum \chi_{E_j} ds$$

$$\le C' \int_\gamma |\tilde{\mathfrak{m}}g|^2 ds \le C''.$$

<u>Section 3</u>. In this section we shall discuss a few results concerning analytic capacity. We first show how to use boundedness of the Cauchy

integral to obtain lower bounds on analytic capacity. (This philosophy is quite old, and was first discussed by Havin. What is different here is only the form of duality which is used.) I learned this particular method from Takafumi Murai, and it is in his book [Mu]. Since it does not seem to be well known, however, it is perhaps worth repeating. Let X be a compact Hausdorff space (e.g. compact subset of $\mathbb{C}$), and let μ be a Borel measure on X (e.g. dxdy). Also let T be a reasonable operator (e.g. the Cauchy integrla with truncated kernel) such that

$$T:L^{\infty}(X),M(X) \to C(X),$$

where M = Borel measure, C = continuous functions. We also require that the adjoint is nice enough to make good sense of $T^{*}\sigma(x)$ and the formal equality

$$\int(Tf)d\sigma = \int(T^{*}\sigma)f\,d\mu,$$

for $\sigma \in M$, $f \in L^{\infty}(X)$.

Lemma 1. <u>Suppose</u> T^{*} <u>satisfies the weak type inequality</u>

$$\mu(\{x:|T^{*}\sigma(x)| > \lambda\}) \leq \frac{M}{\lambda}\|\sigma\|_{M}.$$

<u>Then if</u> $E \subset X$ <u>and</u> $\mu(E) > 0$ <u>there is</u> $F \in K_0 \equiv \{F \in L^{\infty}(E),\ 0 \leq F \leq 1,$ $\int_E F d\mu \geq \frac{\mu(E)}{2}\}$ <u>such that</u> $TF \in L^{\infty}(X)$ <u>and</u>

$$\|TF\|_{L^{\infty}(X)} \leq 2M.$$

The history of this lemma is unclear to me, however Murai says it is not his. See his book for further remarks on the lineage of the lemma. The reader should note that the converse of Lemma 1 is also true; this follows from the proof. The proof of the lemma is fast. Let

$\mathcal{B} = \{g \in C(X): \|g\|_{L^\infty(X)} \le 2M\}$, and let $K = T(K_0)$. Then if the lemma is false, $\mathcal{B} \cap K = \phi$. Since $\mathcal{B}$ and K are convex, by Hahn-Banach there is a $\sigma \in M(X)$ such that

$$\int g d\sigma < \left|\int (Tf) d\sigma\right|$$

for $g \in \mathcal{B}$ and $f \in K_0$. Normalizing $\|\mu\|_M = 1$ we obtain

$$\left|\int (Tf) d\sigma\right| > 2M, \qquad f \in K_0.$$

Now let $E_0 = \{x \in E: |T^*\sigma(x)| \le \frac{2M}{\mu(E)}\}$. By the weak type estimate, $\mu(E_0) \ge \frac{\mu(E)}{2}$, and setting $f = \chi_{E_0}$ we have

$$\left|\int (Tf) d\sigma\right| = \left|\int (T^*\sigma) f \, d\mu\right|$$

$$\le 2M,$$

which is a contradiction.

Now if for example $X = K$ where $K \subset$ Lipschitz curve has $m_1(K) > 0$, and if T = Cauchy integral, then $T: L^2(K) \to L^2(K)$, so by real variable theory T^* is weak $(1,1)$. By Lemma 1 there is $f \in L^\infty(K)$, $0 \le f \le 1$, $\int_K f\, ds \ge 1/2\, m_1(K)$, such that $\|Tf\|_{L^\infty(K)} \le 2M$. But then $Tf \in H^\infty(\mathbb{C}\setminus K)$ and $\|Tf\|_{H^\infty} \le 2M$. Furthermore,

$$Tf'(\infty) = \int_K f\, ds \ge 1/2\, m_1(K),$$

so $\gamma(K) \ge CM^{-1} m_1(K)$.

To study when $\gamma(K) = 0$ we first turn to Garnett's example from [G1]. Let $K_0 \subset \mathbb{R}$ be the Cantor set formed by constant ratio of disection whose first stage construction is to remove $(1/4, 3/4)$ from $[0,1]$. Setting $K = K_0 \times K_0$, we see that $0 < m_1(K) < \infty$, and K is "regular" in the sense of Guy David:

$$m_1(K \cap \{|z-z_0| \le r\}) \sim r$$

for $z_0 \in K$ and $r \le 1$. Garnett showed $\gamma(K) = 0$. By duality (e.g. Lemma 1), the Cauchy integral cannot be bounded on $L^2(K,dm_1)$. David [D2] proved a quantitative form of this to get the lower bound for the Cauchy integral operator in terms of the Lipschitz constant of the curve. We present here a very elementary proof of Garnett's theorem that also gives an estimate (very weak!) for the lower bound of the Cauchy integral.

For $z \in \Omega = \overline{\mathbb{C}}\setminus K$, let $d(z) = \text{distance}(z,K)$. Let $G(z,z_0)$ denote Green's function for Ω with pole at z_0, and let $w_{z_0}(E)$ denote the harmonic measure of $E \subset K$ measured at z_0. We use the following well known estimate, the proof of which is left as an exercise. First some notation. Let $\{Q_j^n\}_{j=1}^{4^n}$ be the squares of sidelength 4^{-n} which contain K and suppose $d(z) \sim 4^{-n}$. Let $Q(z)$ denote a square Q_j^n which is closest to z.

Lemma 2. If $0 < 2d(z) \le d(z_0) \le 1$, then

$$G(z,z_0) \sim w_{z_0}(Q(z)).$$

We use Lemma 1 and the relation

$$w_z(Q(z)) \sim 1$$

to prove Garnett's lemma. Let $f \in H^\infty$, $\|f\|_\infty = 1$, vanish at ∞; we use the Cauchy integral formula to show $f(z) \equiv 0$. Let γ_j^n denote the boundary of the square $3/2\ Q_j^n$. Then $\text{distance}(\gamma_j^n,\gamma_k^m) \ge c4^{-n}$ if $(n,j) \ne (m,k)$. If $m > n$ we use the abusive notation $\gamma_k^m \subset \gamma_j^n$ to mean γ_k^m lies inside γ_j^n. Let $\delta > 0$ be small and let

$$D(\gamma_j^n) = \sup_{z\in\gamma_j^n} 4^{-n}|f'(z)|.$$

Our aim is to prove

Lemma 3. For each γ_j^n there is $\gamma_k^m \subset \gamma_j^n$ such that

$$D(\gamma_k^m) \le c\delta$$

and

$$m \le n + c'\delta^{-2}.$$

Now by iterating Lemma 3 we see we can surround K by a finite number of curves γ_j^n lying in two families, $\mathcal{F}_1$ and $\mathcal{F}_2$. The curves $\gamma_j^n \in \mathcal{F}_1$ satisfy $D(\gamma_j^n) \le c\delta$, and $\mathcal{F}_2$ satisfies

$$\sum_{\gamma_j^n\in\mathcal{F}_2} m_1(\gamma_j^n) \le \delta.$$

Since $\sum_{\gamma_j^n\in\mathcal{F}_1} m_1(\gamma_j^n) \le 6$, we can pick $z_j^n \in \gamma_j^n$ and write

$$|2\pi i\ f(\zeta)| = \left| \sum_{\gamma_j^n\in\mathcal{F}_1} \int_{\gamma_j^n} \frac{f(z)dz}{z-\zeta} + \sum_{\gamma_j^n\in\mathcal{F}_2} \int_{\gamma_j^n} \frac{f(z)dz}{z-\zeta} \right|$$

$$\le C\delta + \sum_{\gamma_j^n\in\mathcal{F}_1} \left\{ \int_{\gamma_j^n} \frac{|f(z)-f(z_j^n)|\,|dz|}{|z-\zeta|} + \left| \int_{\gamma_j^n} \frac{f(z_j^n)dz}{z-\zeta} \right| \right\}$$

$$\le C\delta + \sum_{\gamma_j^n\in\mathcal{F}_1} (C\delta m_1(\gamma_j^n) + 0)$$

$$\le C'\delta,$$

whenever distance $(\zeta,K) \ge 1$.

To prove Lemma 3 we invoke Green's formula to obtain

$$(1)\qquad \iint_\Omega |f'(z)|^2 G(z,z_0)\,dxdy = c\int_K |f(\zeta) - f(z_0)|^2 dw_{z_0}(\zeta)$$

$$\leq C$$

because $\|f\|_\infty \leq 1$. Let A_j^n denote the "annulus" $\{z:\text{distance}(z,\gamma_j^n) \leq 1/8\ 4^{-n}\}$, so that $A_j^n \cap A_k^m = \phi$, $(n,j) \neq (m,k)$. Fixing $z_0 \in \gamma_j^n$ we see from (1) and the inequality

$$D(\gamma_k^m)^2 \leq C\iint_{A_k^m} |f'(z)|^2 dxdy$$

that

$$\sum_{\gamma_k^m \subset \gamma_j^n} D(\gamma_k^m)^2 w_{z_0}(Q_k^m) \leq C \sum_{\gamma_k^m \subset \gamma_j^n} \iint_{A_k^m} |f'(z)|^2 G(z,z_0)\,dxdy$$

$$\leq C'.$$

Now let $\mathcal{F} = \{\gamma_k^m \subset \gamma_j^n : m \leq n + C_0\delta^{-2}\}$. Then $\sum_{\gamma_k^m \in \mathcal{F}} \chi_{Q_k^m} \geq C_0\delta^{-2}$ on $K \cap Q_j^n$.

Suppose the lemma is false, so that $D(\gamma_k^m)^2 \geq \delta$ for $\gamma_k^m \in \mathcal{F}$. Then

$$\sum_{\gamma_k^m \in \mathcal{F}} D(\gamma_k^m)^2 w_{z_0}(Q_k^m)$$

$$\geq \sum_{\mathcal{F}} \delta^2 w_{z_0}(Q_k^m)$$

$$= \delta^2 \int_{K\cap Q_j^n} \sum_{\mathcal{F}} \chi_{Q_j^m} dw_{z_0}$$

$$\geq \delta^2 \cdot C_0\delta^{-2} \cdot w_{z_0}(K\cap Q_j^n)$$

$$\geq cC_0.$$

This contradicts our last inequality if C_0 is large enough.

We remark that this proof gives some (minor) credence to a "conjecture" of Carleson. Define $BMO(\Omega)$ to be the space of all analytic functions on Ω satisfying

$$\sup_{z_0\in\Omega} \iint_{\Omega} |f'(z)|^2 G(z,z_0)\,dxdy \le \|f\|_{BMO(\Omega)} < \infty.$$

Carleson asked if it is true that $H^\infty(\Omega)$ is trivial if and only if $BMO(\Omega)$ is trivial. It seems reasonable to believe that Carleson had found an argument similar to the one just presented.

We now outline the philosophy behind some results due to Murai. For a set $K \subset \mathbb{C}$, $r \in \mathbb{R}$, and $\theta \in [0,\pi]$, let $N(r,\theta)$ denote the number of points in K which lie on the line orthogonal to $\{te^{i\theta} : t \in \mathbb{R}\}$ and passing through $re^{i\theta}$. Then the α Crofton length of K is defined by setting

$$Cr_\alpha(K) = \int_0^\pi \int_{\mathbb{R}} N(r,\theta)^\alpha \,dr d\theta, \qquad \alpha \ge 0$$

where $N(r,\theta)^0$ is defined to be 1 if $N(r,\theta) \ne 0$, and $N(r,\theta)^0 = 0$ if $N(r,\theta) = 0$. Vitushkin had asked in the 1960's if $\gamma(K) = 0 \Leftrightarrow Cr_0(K) = 0$. Mattila [Mat] found an example of a compact K and a Möbius transformation τ such that $Cr_0(K) = 0$ but $Cr_0(\tau(K)) \ne 0$. That disproves Vitushkin's conjecture because $\gamma(K) = 0 \Leftrightarrow \gamma(\tau(K)) = 0$. (This since τ preserves analytic functions.) The amusing quandry is that one then knows the conjecture is false, but one does not know which implication fails! Due to the work of Murai we have a much fuller understanding of the situation.

Theorem 4 (Murai). _If_ $\alpha < 1/2$ _there is_ K _compact such that_ $Cr_\alpha(K) = 0$ _but_ $\gamma(K) > 0$.

The proof of Theorem 4 is quite involved, but the central idea is elegant and easy to explain. Fix $K_0 = [0,1]$ and pick $n_1 \in \mathbb{N}$. Write

$$K_1 = \left\{ \bigcup_{j=0}^{2^{2n_1}-2} \left[2j2^{-2n_1}, (2j+1)2^{-2n_1} \right] + i2^{-2n_1-1} \right\}$$

$$\cup \left\{ \bigcup_{j=0}^{2^{2n_1}} \left[(2j+1)2^{-2n_1}, (2j+2)2^{-2n_1} \right] - i2^{-2n_1-1} \right\} .$$

Then the unit interval has been divided into 2^{2n_1} intervals of length 2^{-2n_1} and these have been alternately pushed up and pulled down so as to be separated by 2^{-2n_1}. Now pick $n_2 \in \mathbb{N}$ and alter K_1 by taking each interval in K_1, scaling it to $[0,1]$, invoking the previous procedure, and then scaling it back. We then obtain a set $K_2 = K(n_1,n_2)$, and we can proceed by induction to build sets $K_M = K(n_1,n_2,\ldots,n_M)$. Murai calls K_M a crank of type $(n_1,\ldots,n_M)$. The idea is that $Cr_0(K_M) \le C_\varepsilon M^{-1+\varepsilon}$, while $\gamma(K_M) \ge CM^{-1/2}$. Because the homogeneities are different $(M^{-1+\varepsilon} << M^{-1/2})$, one can patch together such sets to get the desired counter-example. (However, that part is not trivial.) We outline proofs of the above two inequalities.

The inequality $\gamma(K_M) \ge CM^{-1/2}$ follows from Murai's estimate

(2) $$\|C\|_{L^2(\gamma,ds)} \le C(1 + M)^{1/2}$$

if γ is a Lipschitz graph of Lipschitz constant M. (See e.g. [Mu].) Now by adding line segments of slope M to the endpoints of the intervals in K_1, one sees there is a Lipschitz graph γ_1 of Lipschitz constant M such that $m_1(\gamma_1 \cap K_1) = 1 - \frac{1}{M}$. Repeating this construction, we see that for each n there is γ_n of Lipschitz constant M such that $m_1(\gamma_n \cap K_n) \ge (1 -$

$\frac{1}{M})^n$. Setting $\gamma = \gamma_M$ and $E = \gamma_M \cap K_M$ we have $m_1(E) \sim e^{-1}$. By (2) and Lemma 1 there is $f \in H^\infty(E^c)$ with $\|f\|_{H^\infty} \le CM^{1/2}$ and $f'(\infty) = 1$. Thus $\gamma(K_M) \ge \gamma(E) \ge CM^{-1/2}$.

To obtain the estimate on $Cr_0(K_M)$ we turn to probability theory. First fix a "sample" value of θ, $\theta = \pi/4$. We will "show" that

$$\int_{\mathbb{R}} N(r,\pi/4)^0 dr \le CM^{-1}.$$

That does not entirely prove the desired estimate, but the proof contains the difficulties confronted for "most" θ. Fix a point $x \in K_0 = [0,1]$ and let L be the line of slope one which hits x. Then if n_1 is large, and if P = probability,

$$P(L \cap K_1 = \phi) \simeq 1/2$$

and

$$P(\#(L \cap K_1) = 2) \simeq 1/2.$$

Now if $L \cap K_1 = \phi$ and if n_2 is very large, we expect $L \cap K_2 = \phi$. Therefore, for each fixed $x \in [0,1]$ and $L = L(x)$, we expect the quantity

$$S_n(x) = \#(L \cap K_n)$$

to mimic some absorbing random walk. Let X_n be independent random variables which take on ± 1 with probability $1/2$, and let $Y_n = X_1 + X_2 + \cdots + X_n$. Let Z_n be defined by setting $Z_1 \equiv 1$ and

$$Z_n(x) = Z_{n-1}(x) + Y_{Z_{n-1}(x)}(x), \quad n \ge 1.$$

Then $Z_n \ge 0$ and S_n is much like Z_n. Now the process Z_n is an old quantity, and it is well known that

$$P(Z_n \ne 0) \sim n^{-1}.$$

Converting this estimate back to S_n, we expect that for K_M,

$$\int_{\mathbb{R}} N(r,\pi/4)^0 dr \le \frac{1}{M},$$

which is in fact true. This argument cannot give a good estimate when $|\theta - \pi/2| \le \frac{1}{M}$, but some version works for the other θ, and this will give us $Cr_0(K_M) \le C_\epsilon M^{-1+\epsilon}$.

Section 4. Makarov domains

Let $\Omega \subset \mathbb{C}$ be a simply connected domain, $\Omega \ne \mathbb{C}$, and let $\omega(E)$ denote the harmonic measure of a set $E \subset \partial\Omega$ measured at some fixed point in Ω. A beautiful result due to Makarov [Mak] asserts that ω is supported on a set of (Hausdorff) dimension one and, in a quantifiable sense, the support cannot be smaller than dimension one. For $h \nearrow$ a measure function, define as usual (see e.g. [Mak]) the associated Hausdorff measure Λ_h. Makarov showed that the function $h_C(t) = t \exp\{C\sqrt{\log\frac{1}{t} \log\log\log\frac{1}{t}}\}$, $C > 0$, is the proper measure for the size of support ω.

Theorem A. (Makarov) <u>There is</u> $C_0 > 0$ <u>such that for any simply connected</u> Ω <u>and any</u> $E \subset \partial\Omega$,

$$\Lambda_{h_{C_0}}(E) = 0 \Rightarrow \omega(E) = 0.$$

Theorem B. (Makarov) <u>There is</u> $C_1 > 0$ <u>and an</u> Ω <u>bounded by a Jordan curve</u>, together with $E \subset \partial\Omega$ such that

$$\Lambda_{h_{C_1}}(E) = 0, \quad \omega(E) = 1.$$

The purpose of this section is to investigate when ω is maximally compressed, i.e. when the conclusion of theorem B holds. We say that Ω is a

(C) Makarov domain if there is $E \subset \partial\Omega$ with $\Lambda_{h_C}(E) = 0$, $\omega(E) = 1$.

Makarov exhibited a Makarov domain by writing $\Omega = f(D)$, D = disk, where $\log f'(z) = \sum\sum_{n=1}^{\infty} z^{2^n}$, and then using properties of $\log f'$ discovered by Salem-Zygmund [Sa, Zy] and Mary Weiss [W]. We are interested here in geometric conditions on Ω which guarantee it is a Makarov domain. Our results are inspired by the recent papers of Przytycki et al. [P, U, Zd] and Makarov [Mak] where the following result in iteration theory is proven.

Theorem C. Let $\Omega = A_\infty$ be the domain of attraction at ∞ for the polynomial z^2+C, and suppose that z^2+C acts hyperbolically on its Julia set. Then Ω is a Makarov domain.

From Makarov's original example in theorem B and also from theorem C, we see that for a Makarov domain, $\partial\Omega$ should be wiggly on many scales. For another clue on the tact we should take, we recall that Makarov's theorems A and B depend in a crucial way upon Makarov's Law of the Iterated Logarithm (MLIL):

Theorem D. (Makarov) If $\varphi \in B$, the Bloch space,

$$\overline{\lim_{r\to 1}} \frac{|\varphi(re^{i\theta})|}{\sqrt{\log \frac{1}{1-r} \log_3 \frac{1}{1-r}}} \le C\|\varphi\|_B$$

for a.e. angle θ.

Here we have abbreviated $\log_3 = \log\log\log$. Recall that a function φ holomorphic on D is in the Bloch space if

$$\|\varphi\|_B = \sup_{z\in D} |\varphi'(z)|(1-|z|) < \infty.$$

The space B arises naturally because $\log f' \in B$ if f is univalent on

D. Makarov showed that if

$$(1) \qquad \overline{\lim_{r\to 1}}\ - \frac{\operatorname{Re}\log f'(re^{i\theta})}{\sqrt{\log \frac{1}{1-r}\log_3 \frac{1}{1-r}}} \geq \varepsilon > 0 \quad \text{a.e.}d\theta,$$

then $f(D)$ is an (ε') Makarov domain for all $\varepsilon' < \varepsilon$. Our strategy is therefore to find geometric conditions on Ω so that (1.1) holds; theorem D must enter somewhere in our estimates. Let $\rho(\cdot,\cdot)$ denote the standard hyperbolic metric on D. Our first theorem should be compared with Mary Weiss [W].

Theorem 1. <u>Suppose</u> $\varphi \in B$, $\|\varphi\|_B \leq 1$, <u>and suppose</u>

$$(2) \qquad \sup_{\rho(z,z_0)\leq 1} |\varphi'(z)|(1-|z|) \geq \varepsilon > 0$$

<u>for all</u> $z_0 \in D$. <u>Then</u>

$$\overline{\lim_{r\to 1}}\ \frac{\operatorname{Re}\varphi(re^{i\theta})}{\sqrt{\log \frac{1}{1-r}\log_3 \frac{1}{1-r}}} \geq C(\varepsilon) > 0 \quad \text{a.e.}d\theta.$$

For a domain Ω and $z_0 \in \Omega$ we now define some terminology. Let $z_1 \in \partial\Omega$ be a closest point in the Euclidean topology, let $\theta_0 = \arg(z_0 - z_1)$, and let $\tilde{z}_1 = z_1 + \delta(z_0 - z_1)$ where $0 < \delta < 1$ is fixed. Define

$$L^\delta_{z_0} = \{z_1 + te^{i(\theta_0 - \pi/2)} \quad t \in \mathbb{R},$$
$$|t| \leq \delta^{-1}|z_1 - z_0|\},$$

and

$$S^\delta_{z_0} = \{z | z - z_1| < \delta^{-1}|z_1 - z_0|,$$
$$|\theta_0 - \arg(z - \tilde{z}_1)| < \pi/2\}.$$

Then $L^\delta_{z_0}$ is some sort of "tangent" to $\partial\Omega$ as seen from z_0, and $S^\delta_{z_0}$ is a half-disk more or less lying in Ω. We say that Ω satisfies condition $M(\delta)$ if for all $z_0 \in \Omega$ either

(3) $$\exists z \in L_{z_0}^{\delta}, \ \text{distance}(z,\partial\Omega) \geq \delta|z_1-z_0|$$

or

(4) $$S_{z_0}^{\delta} \cap \partial\Omega \neq \phi.$$

A domain satisfying $M(\delta)$ is evidently wiggly on all scales. For example, it is easy to prove that the domains considered in theorem C satisfy $M(\delta)$ for some $\delta > 0$.

Theorem 2. If Ω satisfies $M(\delta)$, $\log f'$ satisfies (2) for $\varepsilon = \varepsilon(\delta) > 0$, whenever $f:D \xrightarrow{\text{onto}} \Omega$ is a conformal mapping. Conversely, if $\log f'$ satisfies (2), Ω satisfies $M(\delta)$ for $\delta = \delta(\varepsilon) > 0$.

Corollary 3. If Ω satisfies $M(\delta)$, Ω is a $(C(\delta))$ Makarov domain.

The proof of theorem 2 is an elementary and entirely straightforward exercise in normal families; we omit the proof. The remainder of this paper is devoted to the proof of theorem 1. While not short, the proof is again elementary, and can be read by anyone with a passing knowledge of Littlewood-Paley theory and Khinchin's Law of the Iterated Logarithm for Bernoulli Trials. The reader will notice that our proof merely follows the argument in Feller [F], pp. 205, with the necessary adaptations. We remark that our proof of theorem 1 can also be modified so as to be valid in the ball in $\mathbb{C}^n$.

Proof of Theorem 1.

Let $M = M(\varepsilon)$ and $a = a(\varepsilon)$ be positive constants to be fixed later; M is large and a is small. Fix a positive integer n, set $R = (a \log n)^{-1} M^n$, and define

$$S_n(z) = \varphi((1-2^{-M^n})z)$$

and

$$X_k^n(z) = \varphi((1-2^{-kR})z) - \varphi((1-2^{-(k-1)R})z), \quad 1 \le k \le a \log n.$$

Then since we may assume $\varphi(0) = 0$,

$$S_n = \sum_{k=1}^{a \log n} X_k^n .$$

By the definition of φ we also have

$$\|X_k^n\|_B \le 2, \quad |(X_k^n)'(z)| \le 2 \cdot 2^{kR}. \tag{5}$$

Our first observation is the following lemma. By $\omega_z(E)$ we denote the harmonic measure of E at z.

Lemma 3. <u>There are</u> $\alpha = \alpha(\varepsilon)$, $\beta = \beta(\varepsilon) > 0$ <u>such that for any</u> N_0 <u>and any</u> $R > R(N_0)$,

$$\omega_{z_0}(\{\theta : \mathrm{Re}\, X_k^n(e^{i\theta}) \ge 2\alpha\sqrt{R}\}) > \beta$$

<u>for any</u> z_0, $|z_0| < 1-2^{-(k-1)R-N_0}$.

Proof. First let $z_0 = 0$. Then since $X_k^n(0) = 0$, Green's formula yields

$$\int_0^{2\pi} |X_k^n(e^{i\theta})|^2 d\theta = C\iint_0 |(X_k^n)'|^2 \log\frac{1}{|z|}\, dxdy$$

$$\geq C' \int_0^{2\pi} \{\int_{\Gamma_\theta} |(X_k^n)'|^2 dxdy\} d\theta$$

$$= C' \int_0^{2\pi} A^2(X_k^n)(\theta) d\theta.$$

Here we denote by Γ_θ the standard Stolz cone with vertex $e^{i\theta}$ and aperature $2\pi/3$, and set

$$A^2(F)(e^i\theta) = \int_{\Gamma_\theta} |F'|^2 \, dxdy$$

to be the Lusin area function of F. Now by the form of X_k^n together with (5) and (2), there is $C = C(\varepsilon) > 0$ such that for $\tilde{\Gamma}_\theta = \{z \in \Gamma_\theta: r_1 = 1-2^{-(C+(k-1)R)} \leq |z| \leq 1-2^{C-kR} = r_2\}$,

$$A^2(X_k^n)(e^{i\theta}) \geq \int_{\tilde{\Gamma}_\theta} |(X_k^n)'|^2 dxdy$$

$$\geq C \int_{r_1}^{r_2} (\tfrac{\varepsilon}{2}(1-r)^{-1})^2 (1-r) dr$$

$$= C\varepsilon^2 R.$$

Consequently,

$$\|X_k^n(e^{i\theta})\|_{L^2} \geq C\varepsilon R^{1/2} . \tag{6}$$

On the other hand, (1) and the form of X_k^n give

$$A^2(X_k^n)(e^{i\theta}) = \int_{\tilde{\Gamma}_\theta} |(X_k^n)'|^2 + \int_{\Gamma_\theta \setminus \tilde{\Gamma}_\theta} |(X_k^n)'|^2$$

$$\leq C \int_{r_1}^{r_2} (2(1-r)^{-1})^2 (1-r) dr + C$$

$$\leq CR.$$

By the equivalence of L^p norms of functions and their area functions (see e.g. Stein [St]) and since $X_k^n(0) = 0$,

(7) $$\|X_k^n(e^{i\theta})\|_{L^4} < C\|A(X_k^n)(e^{i\theta})\|_{L^4} \le CR^{1/2} .$$

Now since $X_k^n(0) = 0$,

(8) $$\int_0^{2\pi} X_k^n(e^{i\theta})d\theta = 0,$$

while by the fact that the Hilbert transform is unitary on $\{f \in L^2(d\theta): f(0) = 0\}$,

(9) $$\|Re(X_k^n)(e^{i\theta})\|_{L^2} = \|Im(X_k^n)(e^{i\theta})\|_{L^2}.$$

It is an exercise with Chebychev's inequality to verify that functions satisfying (6)-(9) also satisfy the conclusion of lemma 3 for $z_0 = 0$.

To handle the case where $z_0 \neq 0$, set $\tilde{X}_k^n(z) = X_k^n(\tau(z))$, where $\tau: D \to D$ is a Mobius transformation sending 0 to z_0. Then by the form of X_k^n it is immediate that

$$|\tilde{X}_k^n(0)| \le 2N_0.$$

Arguing in exactly the same manner, we see that (6)-(9) hold for $\tilde{X}_k^n(e^{i\theta}) - \tilde{X}_k^n(0)$ once $R \ge R(N_0)$.

Lemma 4. <u>Let</u> $S_n^* = \sum_{k=M^{-1}a \log n+1}^{a \log n} X_k^n$. <u>Then for any</u> z_0 <u>such that</u>

$$|z_0| < 1-2^{-M^{n-1}},$$

$$\omega_{z_0}(\{\theta: ReS_n^*(e^{i\theta}) > a\alpha\sqrt{M^n}\log n\})$$

$$\geq (\frac{\beta}{2})^{a \log n} \geq n^{-1/2} ,$$

<u>if</u> $a \log 2/\beta \leq 1/2$ and $n \geq n_0$.

Proof. Let $E_k = \{\theta: \text{Re } X_k^n(\theta) \geq \alpha\sqrt{R}\}$ and let $\tilde{E}_k = \{\theta: \text{distance}(\theta, E_k) \leq 2^{-kR}\}$. Then by (2.1) $\text{Re} X_k^n \geq 2\alpha\sqrt{R} - 2$ on $\tilde{E}_k$. Let $\tilde{E}_k = \bigcup_j I_j^k$ be the decomposition of $\tilde{E}_k$ into open intervals. Then by lemma 3,

$$|\tilde{E}_{k+1} \cap I_j^k| \geq \frac{2\beta}{3}|I_j^k|$$

for each interval I_j^k as soon as N_0 in lemma 3 is large enough with respect to β. Consequently,

$$|\overset{*}{E}{}^n| = |\bigcap_{k=M^{-1}a \log n+1}^{a \log n} \tilde{E}_k| \geq (\frac{\beta}{2})^{a \log n}$$

if $n \geq n_0 = n_0(\beta)$. Furthermore,

$$\sum_{k=M^{-1}a \log n+1}^{a \log n} X_k^n \geq (\frac{2}{3}\, a \log n)(2\alpha\sqrt{R} - 2)$$

$$\geq a\alpha\sqrt{M^n \log n} \quad \text{on } \overset{*}{E}{}^n. \qquad \blacksquare$$

We now recall a slightly stronger version of Theorem D (MLIL) which is contained in Makarov [Mak].

Lemma 5. <u>If</u> $\|\varphi\|_B \leq 1$, $\varphi(0) = 0$,

$$\|\varphi((1-r)e^{i\theta})\|_{L^{2p}}^{2p} \leq C^p p^p (\log \frac{1}{1-r})^p.$$

Now let $1-r_{n-1} = 2^{-M^{n-1}}$ and let $p = \log_3 \frac{1}{1-r_{n-1}}$. Then by lemma 5 and Chebychev's inequality,

$$(10) \qquad |\{\theta: |\varphi(r_{n-1}e^{i\theta})| \geq A_0\sqrt{M^{n-1}\log n}\}| \leq \frac{1}{n^2}$$

if A_0 is large enough. Since $\varphi(r_n z) = S_{n-1}(z)$ and since $S_n^* = S_n - S_{n-1}$, lemma 4 together with (10) gives us

$$\text{(11)} \qquad \overline{\lim} \frac{\operatorname{Re} S_n(e^{i\theta})}{\sqrt{M^n \log n}} \geq \frac{a\alpha}{2} \qquad \text{a.e. } d\theta$$

as soon as $A_0 M^{-1/2} < \frac{a\alpha}{2}$. This follows from the fact that $\Sigma n^{-1/2} = \infty$ while $\Sigma n^{-2} < \infty$. The details of this Borel-Cantelli argument are a real variables argument best left to the reader. Theorem 1 now follows from (11) because $|S_n(e^{i\theta}) - \varphi(r_n e^{i\theta})| \leq 1$.

Section 5. Problems

We close with some questions.

Problem 1. Suppose $K \subset \Gamma$, $m_1(K) > 0$ where Γ is a Lipschitz curve. Construct by hand a non constant function in $H^\infty(K^c)$.

It has been known since Calderon's theorem [C] that one can find such functions (see e.g. Section 3). However, all proofs use the Hahn-Banach theorem. A construction of such functions might help to prove the corona theorem for $H^\infty(K^c)$, thus generalizing the known case where $\Gamma = \mathbb{R}$. (See Acta Math. 155). I had hoped that the method of Section 1 would solve this problem, but so far all efforts have come to nought.

Problem 2. Let $\Gamma \subset \mathbb{R}^{n+1}$ be a Lipschitz surface and let $BMO(\Gamma)$ denote BMO on Γ with respect to surface measure. Let K_j be the operator corresponding to the kernel $\frac{x_j}{|x|^{n+1}}$. Is it possible to modify Uchiyama's construction [U] to obtain $BMO(\Gamma) \subset L^\infty + \sum_j K_j L^\infty$?

The insights gained from solving this problem might help to solve problem 1.

Problem 3. What role does the "Geometric Lemma" of Section 1 actually play in bounding singular integrals on surfaces in $\mathbb{R}^n$?

Guy David and Stephen Semmes have some recent results on singular integrals on surfaces, but it is not clear if the Geometric Lemma is lurking in the background or is just an artifact of dimension 1.

For our next problem we require an introduction. Suppose we are given a configuration K_n of n disks, all of radius 1, such that K_n is connected, and disks in K_n either osculate or are disjoint. Send a disk (of radius 1) from infinity by Brownian motion until it osculates some disk in K_n. Stop the picture and call it K_{n+1}. This process is well defined, and if K_1 = unit disk we obtain a nice probability distribution of possible configurations for K_n.

Problem 4. Is there a continuous version of this process?

This is the problem of D.L.A. (Diffusion Limited Aggregation), much beloved by certain physicists. Experimental evidence from computer simulations shows that the exterior of K_n is essentially one of the Makarov domains of Section 4. Is Makarov's LIL related to DLA?

REFERENCES

[C] A.P. Calderón, "Cauchy integrals on Lipschitz curves and related topics," Proc. Nat. Acad. Sci. U.S.A. 74(1977), 1324-1327.

[Co,Mc,Me] R. Coifman, A. McIntosh, and Y. Meyer, "L,intégrale de Cauchy définit un opéateur borné sur L^2 pour les courbes lipschitziennes," Ann. Math. 116(1982), 361-387.

[D1] G. David, "Opérateurs integraux singuliers sur certaines courbes du plan complexe." Ann. Scient. Ec. Norm. Sup. 17(1984), 157-189.

[D2] __________, "Une minoration de la norme de l'operateur de Cauchy sur les graphes lipschitziens," to appear T.A.M.S.

[D,J] G. David and J.L. Journé, "A boundedness criterion for general Calderón-Zygmund operators," Ann. Math. 120(1984), 371-397.

[D,J.Se] G. David, J.L. Journé, and S. Semmes, Opérateurs de Calderón-Zygmund, fonctions para acrétives et interpolation," Revista Matemática Iberoamericana 4(1985), 1-56.

[F] W. Feller, An Introduction to probability theory and its applications, Vol. 1, John Wiley and Sons, Inc., 1968.

[G1] J. Garnett, "Positive length but zero analytic capacity," P.A.M.S. 24(1970), 696-699.

[G2] __________, Bounded analytic functions, Academic Press, 1981.

[Je,K] D. Jerison and C. Kenig, "Hardy spaces, A_∞, and singular integrals on chord-arc domains," Math. Scand. 50(1982), 221-247.

[Jo,Se] P.W. Jones and S. Semmes, "An elementary proof of the boundedness of Cauchy integrals on Lipschitz curves," preprint.

[K] C. Kenig, "Weighted H^p spaces on Lipschitz domains," Amer. J. Math. 102(1980), 129-163.

[U] A. Uchiyama, "A constructive proof of the Fefferman-Stein decomposition of BMO(R^n)," Acta Math. 148(1982), 215-241.

[W] M. Weiss, "The law of the iterated logarithm for lacunary trigonometric series, T.A.M.S. 91(1959), 444-469.

[Z] M. Zinsmeister, "Domaines reguliers du plan," Ann. Inst. Fourier, Grenoble, 35(1985), 49-55.

[Mak] N.G. Makarov, "On the distortion of boundary sets under conformal mappings," Proc. London Math. Soc. 51(1985), 369-384.

[Mat] P. Mattila, "Smooth maps, null sets for integral geometric measure and analytic capacity," Ann. Math. 123(1986), 303-309.

[Mu] T. Murai, <u>A real variable method for the Cauchy transform and applications to analytic capacity</u>, to appear in Springer Lecture Notes series.

[P,Ur,Zd] F. Przytyeki, M. Urbanski, A. Zdunik, Harmonic, Hausdorff, and Gibbs measures on repellers for holomorphic maps, preprint 1986.

[Sa,Zy] R. Salem and A. Zygmund, "La loi du logarithme itéré pour les séries trigonometriques lacunaires," Bull. Sci. Math. 74(1950), 209-224.

[St] E.M. Stein, <u>Singular integrals and differentiability properties of functions</u>, Princeton University Press, 1970.

RESTRICTION THEOREMS, CARLEMAN ESTIMATES, UNIFORM SOBOLEV INEQUALITIES AND UNIQUE CONTINUATION

BY CARLOS E. KENIG*
DEPARTMENT OF MATHEMATICS
UNIVERSITY OF CHICAGO
CHICAGO, ILLINOIS 60637

Acknowledgement. This article is based on lectures presented at the El Escorial Seminar in Harmonic Analysis and P.D.E., El Escorial, Spain, June 1987. I would like to thank the organizing committee for their invitation and their warm hospitality, and the participants of the Seminar for their enthusiasm, encouragement and endurance.

Introduction.

The aim of this paper is to attempt to describe the connection between unique continuation theorems for partial differential operators, Carleman estimates, uniform Sobolev inequalities and restriction theorems for the Fourier transform. It is an expanded and updated version of [30].

If a linear partial differential operator $P(x, D)$ has the property that whenever $\Omega \subset \mathbb{R}^n$ is open, connected, and $P(x, D)u = 0$ in Ω implies that if u vanishes of infinite order at $x_0 \in \Omega$, then u must be identically 0 in Ω, we say that $P(x, D)$ has the strong unique continuation property (s.u.c.p.). If $u \equiv 0$ in $\Omega' \subset \Omega$, Ω' open implies that u must be identically 0 in Ω we say that $P(x, D)$ has the unique continuation property (u.c.p.). If support $(u) \subset K$, $K \subset \Omega$, K compact implies that u is identically 0 in Ω, we say that $P(x, D)$ has the weak unique continuation property (w.u.c.p.). These concepts also have obvious extensions to solutions of differential inequalities.

As it is well known, linear elliptic operators with real analytic coefficients have the (s.u.c.p.). Through the work of Hadamard ([21]) on the uniqueness for the Cauchy problem for non-linear differential equations, it became clear that it would be desirable to establish the unique continuation property (which in many cases is equivalent to uniqueness in the Cauchy problem) for operators whose coefficients are not necessarily real analytic, or even C^∞.

The first such unique continuation results are due to T. Carleman ([11]) (1939). He showed that $\Delta + V(x)$ in $\mathbb{R}^2$, with $V \in L^\infty_{\text{loc}}(\mathbb{R}^2)$ has the (s.u.c.p.). To do so, he introduced an idea, the so called Carleman estimates, which permeates the subsequent work in the subject. In this context, Carleman estimates are inequalities of the form

$$\|e^{\lambda\phi(x)} f\|_{L^2(U)} \le C\|e^{\lambda\phi(x)} \Delta f\|_{L^p(U)},$$

*Supported in part by the NSF and the J. S. Guggenheim Foundation.

for $f \in C_0^\infty(U)$, $U \subset \mathbb{R}^2$, bounded, open, for suitable $\phi(x)$, and C independent of λ, for λ tending to ∞. These inequalities easily yield the above mentioned result of Carleman's.

C. Müller ([35]) extended this result to $\mathbb{R}^n$, $n \geq 3$. The late 50's and early 60's saw a great deal of activity in this area and related uniqueness questions. Notable contributions were made by Cordes ([14]), Aronszajn ([4]), Nirenberg ([36]), Calderón ([10]), Hörmander ([22],[23]), Agmon ([1]) and Aronszajn, Krzywicki and Szarski ([5]). Their result was the strongest unique continuation theorem for second order elliptic operators. They showed that if $(a_{jk}(x))$ is a real, $n \times n$ symmetric, positive definite, Lipschitz continuous matrix, then the differential inequality

$$\left| \sum_{j,k} a_{jk}(x) \frac{\partial^2 u}{\partial x_i \partial x_j}(x) \right| \leq \sum_{|\alpha| \leq 1} C_\alpha(x) \left| \frac{\partial^\alpha u}{\partial x^\alpha}(x) \right| \tag{*}$$

has the (s.u.c.p.), when $C_\alpha \in L^\infty_{\text{loc}}(\mathbb{R}^n)$. This was shown by means of an appropriate Carleman estimate. An example of Pliš ([37]) showed that the regularity assumption on $(a_{jk}(x))$ is optimal. Recently, N. Garofalo and F. H. Lin ([18]) have found a new approach to this result, without using Carleman estimates. They show that if u verifies (*), then $|u|$ is locally an A_∞ weight ([34]), and hence cannot vanish of infinite order.

In recent years there has been interest in establishing unique continuation results for solutions of inequalities such as (*), with L^p conditions, $p < \infty$ on the $C_\alpha(x)$, or other conditions on $C_\alpha(x)$ which do not guarantee local boundedness. The reason for this comes from uniqueness questions in non-linear partial differential equations (as we will illustrate in the body of the paper), as well as from some problems in mathematical physics. In fact, if one considers the Schrödinger operator $H = -\Delta + V(x)$ as a self-adjoint operator on $L^2(\mathbb{R}^n)$, one wants to rule out positive eigenvalues (i.e. embedded eigenvalues). The potentials V that arise in quantum physics need not be smooth, or even locally bounded. Moreover, a well-known example of Wigner and Von Neumann ([57]) shows that even when $n = 1$, one can have embedded eigenvalues, for some V. In the 60's, Kato ([29]), Agmon ([2]), B. Simon ([44]) and others developed a philosophy to rule out positive eigenvalues for H. One of the main steps is to establish the (w.u.c.p.) for $-\Delta + V(x)$. Similar questions for $D + \vec{V}(x)$, where D is the Dirac operator, and $\vec{V}$ a vector valued potential also arise (see [9]). In the late 70's and early 80's, a number of results in this direction were obtained by Bérthier ([8]), Georgescu ([20]), Schechter and B. Simon ([42]), Saut and Scheurer ([39]), Amrein, Bérthier and Georgescu ([3]), K. Senator ([43]) and Hörmander ([26]).

Amrein, Bérthier and Georgescu (1981) ([3]) proved the (s.u.c.p.) for $-\Delta + V(x)$, when $V \in L^p_{\text{loc}}(\mathbb{R}^n)$, $p > n/2$, for $n = 2, 3, 4$ (this was improved to $p = n/2$, $n = 3, 4$ by K. Senator ([43]) (1983)), while for larger n the best results were those of L. Hörmander (1983) ([26]), who established the (u.c.p.) for $V \in L^p_{\text{loc}}(\mathbb{R}^n)$, $p \geq 4n - 2/7$, $n > 4$. Hörmander's results also applied to variable coefficient leading terms, and L_{loc}^q first order terms, $q > 3n - 2/2$.

To begin the description of more recent results on the subject, consider the following example.

EXAMPLE 0.1: Let $u(x) = \exp\{-\log 1/|x|\}^{1+\varepsilon}$, $\varepsilon > 0$, and let $V(x) = \Delta u(x)/u(x)$. u vanishes at 0 of ∞ order, but $V(x) \sim (\log 1/|x|)^{2\varepsilon} \frac{1}{|x|^2} \in L_{loc}^p(\mathbb{R}^n)$ for all $p < n/2$. Thus, (s.u.c.p.) cannot hold for $p < n/2$.

In 1984, D. Jerison and C. Kenig ([28]) showed that $-\Delta + V(x)$ has (s.u.c.p.), for $V \in L_{loc}^{n/2}(\mathbb{R}^n)$, $n \geq 3$. This was done by proving the following Carleman estimate:

Let $n \geq 3$, $\frac{1}{p} - \frac{1}{p'} = \frac{2}{n}$, $\frac{1}{p} + \frac{1}{p'} = 1$, $\tau \notin \mathbb{N} + \frac{n}{p'}$, $\delta = \text{dist}(\tau, \mathbb{N} + n/p')$. Then, there exists $C = C(\delta, n)$, such that, for all $u \in C_0^\infty(\mathbb{R}^n \setminus 0)$

$$\||x|^{-\tau} u\|_{L^{p'}(\mathbb{R}^n)} \leq C \||x|^{-\tau} \Delta u\|_{L^p(\mathbb{R}^n)}. \tag{0.2}$$

Shortly afterwords, E. Stein ([52]) simplified one step in the proof of (0.2), and sharpened it to the Lorentz space inequality

$$\||x|^{-\tau} u\|_{L^{p', p}(\mathbb{R}^n)} \leq C \||x|^{-\tau} \Delta u\|_{L^p(\mathbb{R}^n)}. \tag{0.3}$$

(0.3) in turn implied the (s.u.c.p.) for potentials V in $L_{loc}^{n/2, \infty}(\mathbb{R}^n)$, with the additional hypothesis that

$$\limsup_{\rho \to 0} \|V\|_{L^{n/2, \infty}(B(x_0, \rho))} \leq \varepsilon_n \tag{0.4}$$

for all $x_0 \in \Omega$, and some $\varepsilon_n > 0$, which depends only on the dimension. The proof of (0.2) and (0.3) used complex interpolation in a manner reminiscent of the proof of the L^2 restriction theorem to the sphere, for the Fourier transform ([56]). This theme is one of the main ones in subsequent work.

Also in 1984, E. Sawyer ([41]) established the (s.u.c.p.) for a different class of unbounded potentials in $\mathbb{R}^3$. We say that $V \in K_n^{loc}$ if $\lim_{r \to 0} \sup_{x \in K} \int_{|x-y| < r} \frac{|V(y)|}{|x-y|^{n-2}} dy = 0$, for all $K \subset\subset \Omega$. Notice that $L_{loc}^p \subset K_n^{loc}$ if $p > n/2$, while $L_{loc}^{n/2}$ and K_n^{loc} are not comparable for $n \geq 3$. Sawyer showed that, for $n = 3$, if $V \in K_n^{loc}$, $-\Delta + V(x)$ has (s.u.c.p.). His proof involves also an appropriate Carleman estimate, and the main step in his proof is in fact the same as Stein's simplification of the original Jerison–Kenig argument.

Recently ([12]), S. Chanillo and E. Sawyer have taken up these questions in connection with another class of potentials, introduced by C. Fefferman and Phong ([17]) in their study of the number of eigenvalues of $-\Delta + V(x)$. We say that $V \in J_p^{loc}$ (this terminology is not standard, and has only been adopted for the purposes of exposition) if $\lim_{r \to 0} \sup_{x \in K} \left(\frac{1}{|B_r|} \int_{B_r} |V|^p \right)^{1/p} r^2 = 0$, where B_r is a ball centered at x, of radius r. Using the same complex interpolation idea as in the proof of (0.2), together with the main estimate in Stein's proof of (0.3) (or Sawyer's main estimate in [41]), and some weighted norm inequalities for fractional integrals as in [17], [13] or [33], Chanillo and Sawyer have shown that, for $n \geq 3$, if $V \in J_p^{\text{loc}}$, $p > n - 1/2$, then $-\Delta + V(x)$ has (s.u.c.p.). Notice that $L_{\text{loc}}^{n/2} \subset J_p^{\text{loc}}$, for all $p < n/2$, and that even $L_{\text{loc}}^{n/2,\infty}$, together with (0.4) verifies this. Thus, this constitutes a generalization of the results of Jerison–Kenig and Stein. To compare it with Sawyer's result, note that $K_n^{\text{loc}} \subset J_1^{\text{loc}}$, and that $J_p^{\text{loc}} \subset K_n^{\text{loc}}$ for $p > n/2$.

In 1985, D. Jerison ([27]), gave an alternative proof of (0.2), and was also able to give unique continuation results for $D + \vec{V}(x)$, where D is the Dirac operator. His idea was to employ a discrete version of the L^2 restriction theorem to spheres, for the Fourier transform on $\mathbb{R}^{n-1}$. This is the $L^p \to L^{p'}$ norm estimate for the projection operators associated to spherical harmonic expansions of functions on S^{n-1}, due to C. Sogge ([46]). Thus, if $\xi_k : L^2(S^{n-1}) \to L^2(S^{n-1})$ is the projection operator on spherical harmonics of degree k, Sogge showed that

$$\|\xi_k g\|_{L^{p'}(S^{n-1})} \leq C k^{1-2/n} \|g\|_{L^p(S^{n-1})}, \tag{0.5}$$

where p, p' are as in (0.2).

For future reference, note that this is a result about the projection operators associated with the eigenfunction expansion of the spherical Laplacian Δ_S, since, if P_k is a spherical harmonic of degree k, $\Delta_S P_k = k(k + n - 2)P_k$.

It is not hard to show (and we will do this later), that to prove the (w.u.c.p.) for $-\Delta + V(x)$, $V(x) \in L^{n/2}(\mathbb{R}^n)$, it suffices to prove the Carleman estimate

$$\|e^{-\tau x_n} u\|_{L^{p'}(\mathbb{R}^n)} \leq C \|e^{-\tau x_n} \Delta u\|_{L^p(\mathbb{R}^n)}. \tag{0.6}$$

If one sets $v = e^{-\tau x_n} u$, (0.6) is equivalent to

$$\|v\|_{L^{p'}(\mathbb{R}^n)} \leq C \|[\Delta - 2\tau \frac{\partial}{\partial x_n} + \tau^2] v\|_{L^p(\mathbb{R}^n)}. \tag{0.7}$$

One is then led to the idea of proving Sobolev estimates for second order, constant coefficient elliptic operators, uniformly in the lower order terms. This was done by C. Kenig, A. Ruiz and C. Sogge ([31]). The estimate is

$$\|v\|_{L^{p'}(\mathbb{R}^n)} \leq C \|[\Delta + \vec{a} \cdot \nabla + b] v\|_{L^p(\mathbb{R}^n)}, \tag{0.8}$$

where, p, p' are as in (0.2), $\vec{a} \in \mathbb{C}^n$, $b \in \mathbb{C}$, and C is independent of $\vec{a}$ and b. The inequalities corresponding to (0.8), with p' replaced by s, p replaced by r, with $\frac{1}{r} - \frac{1}{s} = \frac{2}{n}$, and the optimal range of exponents r and s were also obtained in [31]. The main tool in the proof of (0.8) is the so-called Stein–Tomas Lemma ([56]) for $\mathbb{R}^n$, which is the main tool in the L^2 restriction theorems. The Stein–Tomas inequality is

$$\left\| \int_{S^{n-1}} \hat{f}(w) e^{ix\cdot w} dw \right\|_{L^s(\mathbb{R}^n)} \leq C \|f\|_{L^r(\mathbb{R}^n)}. \tag{0.9}$$

It is interesting to note, as was done in [31], that one can go full circle, and obtain (0.9) from the $L^r - L^s$ version of (0.8). In fact, if $\vec{a} = 0$, $b = 1 + i\varepsilon$, the $L^r - L^s$ version of (0.8) is equivalent to the multiplier inequality $\left\| \left\{ \frac{\hat{f}(\xi)}{1 - |\xi|^2 + i\varepsilon} \right\}^{\vee} \right\|_{L^s(\mathbb{R}^n)} \leq C\|f\|_{L^r(\mathbb{R}^n)}$. The imaginary part of the multiplier is $\frac{\varepsilon}{(1-|\xi|^2)^2 + \varepsilon^2}$, and $\frac{\varepsilon}{(1-|\xi|^2)^2+\varepsilon^2} \underset{\varepsilon \to 0}{\rightharpoonup} c\,dw$, and thus (0.9) follows.

It turns out that (0.7) also extends to the case when the leading term is not necessarily elliptic. In fact, let $P(D)$ be a complex constant coefficient second order operator, whose principal part is $Q(D)$, where $Q(\xi) = -\xi_1^2 - \xi_2^2 - \xi_j^2 + \xi_{j+1}^2 + \cdots + \xi_n^2$. Kenig, Ruiz and Sogge ([31]) also proved

$$\|v\|_{L^{p'}(\mathbb{R}^n)} \leq C\|P(D)v\|_{L^p(\mathbb{R}^n)}, \tag{0.10}$$

and Kenig and Sogge ([32]) also showed a version of this inequality for Schrödinger operators $i\frac{\partial}{\partial t} + \Delta$:

$$\|u(x,t)\|_{L^{p'}(\mathbb{R}^{n+1})} \leq C\|[i\frac{\partial}{\partial t} + \Delta + \vec{a}\cdot\nabla_x + b]u(x,t)\|_{L^p(\mathbb{R}^{n+1})} \tag{0.11}$$

where here $\frac{1}{p} + \frac{1}{p'} = 1$, $\frac{1}{p} - \frac{1}{p'} = \frac{2}{n+2}$.

(0.10) and (0.11) are proved using Strichartz' version ([55]) of the Stein–Tomas lemma, for general quadratic surfaces. (0.10) and (0.11) lead to unique continuation theorems for operators whose top order terms are not necessarily elliptic, and include, for instance, the wave operator $\Box = \frac{\partial^2}{\partial x_1^2} - \frac{\partial^2}{\partial x_2^2} - \cdots - \frac{\partial^2}{\partial x_n^2}$. For example, if p is as in (0.2), $\left(\frac{\partial^\alpha u}{\partial x^\alpha}\right)_{|\alpha|=2} \in L^p(\mathbb{R}^n)$, $|P(D)u| \leq |V(x)u(x)|$, where P is as in (0.10), $V \in L^{n/2}(\mathbb{R}^n)$ and u vanishes on one side of a hyperplane, then $u \equiv 0$. It is known ([25]) that if the hyperplane is characteristic for $P(D)$ (i.e. $Q(N) = 0$, where N is the normal to the hyperplane), there exists $u \in C^\infty(\mathbb{R}^n)$, $u \not\equiv 0$, vanishing on one side of the hyperplane, with $P(D)u = 0$. However, of course, $\left(\frac{\partial^\alpha u}{\partial x^\alpha}\right)_{|\alpha|=2} \notin L^p(\mathbb{R}^n)$.

Finally, to come back once more to restriction theorems, the realization that (0.8) implies (0.9), lead C. Sogge ([47]), to an extension of his discrete restriction lemma (0.5), to arbitrary compact manifolds.

Let M be a smooth compact connected manifold of dimension $n - 1 \geq 2$ (to be consistent with (0.5)). Let P be a second order elliptic operator on M, with smooth coefficients and real principal part, self-adjoint with respect to a smooth density $d\mu$. Let $\{\lambda_j\}_{j=0}^\infty$ be its eigenvalues, ordered increasingly. Let ξ_k be the projection operators associated to the sum of the eigenspaces $\sum_{j \in \Lambda_k} \mathcal{H}_j$, where $\Lambda_k = \{j : \sqrt{\lambda_j} \in [k-1, k)\}$. Then, for p, p' as in (0.2), we have

$$\|\xi_k g\|_{L^{p'}(M)} \leq Ck^{1-2/n}\|g\|_{L^p(M)}, \tag{0.12}$$

just as in (0.5).

The main tools used in [47] to establish (0.12) are the following uniform Sobolev inequalities

$$\|u\|_{L^{p'}(M)} \leq Ck^{\sigma-1}\|(P-k^2)u\|_{L^2(M)} + \tag{0.13}$$
$$+ Ck^{\sigma}\|u\|_{L^2(M)},$$

where $\sigma = n - 2/2n$.

It turns out ([47]) that it is possible to establish (0.13) using local coordinates (and in this lies the advantage over the "global" estimate (0.12)), using the Hadamard parametrix ([25], Vol III 17.4) and the Carleson–Sjölin method ([51]).

As was first pointed out by C. Fefferman ([16]), restriction theorems imply results for Bochner–Riesz summability. Using (0.12) and ideas coming from Hörmander's proof of the Weyl formula ([24]), C. Sogge was also able to extend the results known for $\mathbb{R}^n$, $n \geq 3$ on Bochner-Riesz summability ([16],[56]), to arbitrary compact manifolds ([48]).

In the rest of this paper we will try to give a sketch of the proofs of some of the results described above.

Section 1: Strong unique continuation.

We will start out by sketching the proof of the unique continuation theorem in [28]. As mentioned in the introduction, the main step is the Carleman estimate (0.2). Let us first illustrate the mechanism which allows one to pass from Carleman estimates to unique continuation theorems.

By simple limiting arguments (see [28] for the details), it is easy to see that (0.2) extends to $u \in W^{2,p}(\mathbb{R}^n)$ which vanish (in a sense made precise in [28]) of infinite order at the origin. Here, $W^{2,p}(\mathbb{R}^n)$ is the space of functions having two derivatives in $L^p(\mathbb{R}^n)$. Let us now assume that Ω is open, connected, $0 \in \Omega$, $u \in W^{2,p}_{\text{loc}}(\Omega)$, $|\Delta u| \leq |Vu|$, $V \in L^{n/2}_{\text{loc}}(\Omega)$, p is as in (0.2), and u vanishes of infinite order at 0. We want to show that $u \equiv 0$ in Ω. Without loss of generality, we can assume that $B = \{|x| < 1\} \subset \Omega$. Also, choose $\tau \to +\infty$, with $\delta = \operatorname{dist}(\tau, N + \frac{n}{p'})$ constant, > 0. Lets now choose $\rho > 0$, $\rho < 1/2$ so that $C\|V\|_{L^{n/2}(|x|<\rho)} < 1/2$, where C is the constant in (0.2). We will show that $u \equiv 0$ in $\{|x| < \rho\} = B_\rho$, which of course, implies our result.

Let $\varphi \in C_0^\infty(B)$, $0 \le \varphi \le 1$, $\varphi \equiv 1$ for $|x| < 1/2$, and let $f = \varphi u$. We will apply (0.2) to f

$$\begin{aligned}\||x|^{-\tau} u\|_{L^{p'}(|x|<\rho)} &\le C\||x|^{-\tau}\Delta f\|_{L^p(\mathbb{R}^n)} \le \\ &\le C\||x|^{-\tau}\Delta u\|_{L^p(|x|<\rho)} + C\||x|^{-\tau}\Delta f\|_{L^p(|x|\ge\rho)} \le \\ &\le C\||x|^{-\tau}Vu\|_{L^p(|x|<\rho)} + C\||x|^{-\tau}\Delta f\|_{L^p(|x|\ge\rho)} \le \\ &\le C\|V\|_{L^{n/2}(|x|<\rho)}\cdot\||x|^{-\tau}u\|_{L^{p'}(|x|<\rho)} + C\||x|^{-\tau}\Delta f\|_{L^p(|x|\ge\rho)},\end{aligned}$$

by Hölder's inequality, since $\dfrac{1}{p} - \dfrac{1}{p'} = \dfrac{2}{n}$. By our choice of ρ, we conclude that

$$\||x|^{-\tau}u\|_{L^{p'}(|x|<\rho)} \le 2C\||x|^{-\tau}\Delta f\|_{L^p(|x|\ge\rho)},$$

and hence,

$$\left\|\left(\frac{|x|}{\rho}\right)^{-\tau} u\right\|_{L^{p'}(|x|<\rho)} \le 2C\|\Delta f\|_{L^p(|x|\ge\rho)}.$$

Letting $\tau \to +\infty$, we see that $u \equiv 0$ in B_ρ.

Let us now pass to the proof of (0.2). It is convenient to restate (0.2) in the following manner. Let $n \ge 3$, $\dfrac{1}{p} - \dfrac{1}{p'} = \dfrac{2}{n}$, $\dfrac{1}{p} + \dfrac{1}{p'} = 1$, $t \notin \mathbb{N}$, $\delta =$ dist$(t, \mathbb{N})$. Then, there exists $C = C(\delta, n)$ such that, for all $f \in C_0^\infty(\mathbb{R}^n\backslash 0)$,

$$\||x|^{-t}f\|_{L^{p'}(\mathbb{R}^n, \frac{dx}{|x|^n})} \le C\||x|^{-t+2}\Delta f\|_{L^p(\mathbb{R}^n, \frac{dx}{|x|^n})}. \tag{1.1}$$

(1.1) was proved in [28] by complex interpolation.

Recall that $(\Delta^{-z/2}f)^\wedge(\xi) = (2\pi|\xi|)^{-z}\hat{f}(\xi)$, or alternatively, for $0 <$ Re $z < n$,

$$\Delta^{-z/2}f(x) = C_z\int_{\mathbb{R}^n} |x-y|^{-n+z}f(y)dy,$$

where $C_z = \pi^{-n/2}2^{-z}\Gamma\left(\frac{n}{2} - \frac{z}{2}\right)/\Gamma(z/2)$, and moreover, the function $\Delta^{-z/2}f(x)/\Gamma\left(\frac{n}{2}-\frac{z}{2}\right)$ is an entire function of z (see [19] and [50]).

Fix t, and find a positive integer m so that $m - 1 < t < m$. For $0 <$ Re $z < n$, let

$$H_z(x,y) = |x|^{-t}|y|^{n+t-z}C_z\left[|x-y|^{-n+z} - \sum_{j=0}^{m-1} a_{j,z}(x,y)\right],$$

where $a_{j,z}(x,y) = \frac{1}{j!}\left(\frac{\partial}{\partial s}\right)^j |sx - y|^{-n+z}|_{s=0}$, and let $K_z(x,y) = H_z(x,y)/\Gamma\left(\frac{n}{2}-\frac{z}{2}\right)$ to obtain an entire function of z. Let $T_zg(x) = \int_{\mathbb{R}^n} K_z(x,y)g(z)|y|^{-n}dy$. We claim that

$$\|T_2g\|_{L^{p'}(\mathbb{R}^n, dx/|x|^n)} \le C\|g\|_{L^p(\mathbb{R}^n, dx/|x|^n)}. \tag{1.2}$$

It is easy to see that (1.2) implies (1.1). In fact, if $f \in C_0^\infty(\mathbb{R}^n \backslash 0)$, $g = |x|^{-t+2}\Delta f$ also belongs to $C_0^\infty(\mathbb{R}^n \backslash 0)$, and $T_2 g(x) = |x|^{-t} f(x)/\Gamma\left(\frac{n}{2} - \frac{1}{2}\right)$.

(1.2) follows, by complex interpolation ([**49**]) from the following two lemmas.

LEMMA 1.3 ([**28**]). *There exists $C = C(\delta, n)$ such that, for every $\gamma \in \mathbb{R}$, we have*

$$\|T_{i\gamma} g\|_{L^2(\mathbb{R}^n, dx/|x|^n)} \le C e^{C|\gamma|} \|g\|_{L^2(\mathbb{R}^n, dx/|x|^n)}.$$

LEMMA 1.4 ([**52**]). *There exists $C = C(\delta, n)$ such that, for every $\gamma \in \mathbb{R}$, $z = \alpha + i\gamma$, $n - 1 < \alpha < n$, $\frac{1}{r} = \frac{1}{q} - \frac{\alpha}{n}$, $1 < q < \frac{n}{\alpha}$, we have*

$$\|T_z g\|_{L^r(\mathbb{R}^n, dx/|x|^n)} \le C e^{C|\gamma|} \|g\|_{L^q(\mathbb{R}^n, dx/|x|^n)}.$$

Also, real interpolation ([**54**]), using Lemmas 1.3 and 1.4 also gives (0.3).

SKETCH OF THE PROOF OF LEMMA 1.3: For a function h in $C_0^\infty(\mathbb{R}_+)$, we define its Mellin transform $\tilde{h}(\eta) = \int_0^\infty h(r) r^{-i\eta - 1}\, dr$. The change of variables $x = \log r$, $\xi = \eta/2\pi$ yields, via Plancharel's theorem, the identity $\int_0^\infty |h(r)|^2 \frac{dr}{r} = \frac{1}{2\pi} \int_{-\infty}^{+\infty} |\tilde{h}(\eta)|^2 d\eta$. Introduce polar coordinates $r = |x|$, $w = \frac{x}{|x|}$, and write a function $f(x)$ as $f(rw)$. We also define $\tilde{f}(r, w) = \int_0^\infty f(rw) r^{-i\eta-1}\, dr$. In order to prove Lemma 1.3, one calculates the action of T_z on radial functions times spherical harmonics, and takes the Mellin transforms. The key calculation is the following: Let $g \in C_0^\infty(\mathbb{R}_+)$. Then,

$$T_z(g(r) P_k(w)) = T_{z,k}(g) P_k(w),$$

where $P_k(w)$ is a spherical harmonic of degree k, and

$$T_{z,k}(g)^\sim(\eta) = \frac{2^{-z}}{\Gamma\left(\frac{n-z}{2}\right)} \cdot \frac{\Gamma\left(\frac{1}{2}(k - t - i\eta)\right) \cdot \Gamma\left(\frac{1}{2}(n + k + t - z + i\eta)\right)}{\Gamma\left(\frac{1}{2}(k - t + z - i\eta)\right) \cdot \Gamma\left(\frac{1}{2}(n + k + t + i\eta)\right)} \cdot \tilde{g}(\eta).$$

Using Stirling's formula in the form

$$|\log \Gamma(\varsigma) - [(\varsigma - 1/2) \log \varsigma - \varsigma]| \le C \text{ for } \operatorname{Re} \varsigma \ge \delta,$$

and the functional equation $\Gamma(\varsigma)\Gamma(1-\varsigma) = \frac{\pi}{\sin \pi\varsigma}$, one can easily show that, if $\delta = \operatorname{dist}(t, \mathbb{N})$, $z = i\gamma$,

$$\left| \frac{\Gamma\left(\frac{1}{2}(k - t - i\eta)\right) \cdot \Gamma\left(\frac{1}{2}(n + k + t - i\gamma + i\eta)\right)}{\Gamma\left(\frac{1}{2}(k - t + i\gamma - i\eta)\right) \cdot \Gamma\left(\frac{1}{2}(n + k + t + i\eta)\right)} \right| \le C e^{C|\gamma|},$$

where $C = C(\delta, n)$.

Now, for $f \in C_0^\infty(\mathbb{R}^n \backslash 0)$ write $f(rw) = \sum_{k=0}^{\infty} f_k(r) P_k(w)$, where $g_k \in C_0^\infty(\mathbb{R}_+)$ and $P_k(w)$ is a spherical harmonic of degree k, normalized so that $\int_{S^{n-1}} |P_k(w)|^2 dw = 1$. By orthogonality of spherical harmonics of different degrees, we have $\int_{\mathbb{R}^n} |f(x)|^2 \frac{dx}{|x|^n} = \sum_{k=0}^{\infty} \int_0^\infty |f_k(r)|^2 \frac{dr}{r}$. Thus,

$$\int_{\mathbb{R}^n} |T_{i\gamma} f(x)|^2 \frac{dx}{|x|^n} = \sum_{k=0}^{\infty} \int_0^\infty |T_{i\gamma,k}(f_k)(r)|^2 \frac{dr}{r} =$$

$$= \sum_{k=0}^{\infty} \frac{1}{2\pi} \int_{-\infty}^{+\infty} |T_{i\gamma,k}(f_k)^\sim(\eta)|^2 d\eta \leq C e^{C|\gamma|} \sum_{k=0}^{\infty} \int_{-\infty}^{+\infty} |\tilde{f}_k(\eta)|^2 d\eta =$$

$$= C e^{C|\gamma|} \int_{\mathbb{R}^n} |f(x)|^2 \frac{dx}{|x|^n},$$

and 1.3 follows.

SKETCH OF THE PROOF OF LEMMA 1.4: Lemma 1.4 is a consequence of a pointwise bound for the kernel $K_z(x,y)$ of T_z, valid when $n-1 <$ Re $z < n$. This bound is the following

$$|K_z(x,y)| \leq C_z |x|^\gamma |y|^\beta |x-y|^{-n+\alpha},$$

where $\alpha =$ Re z, $\gamma = m - 1 - t + (n - \alpha)$, $\beta = t - (m-1)$. Once one has this pointwise bound, Lemma 1.4 follows from well-known fractional integral estimates in [53].

The bound above is, in turn a consequence of a result about ultraspherical polynomials.

LEMMA 1.5 ([52]). *Let* $(1 - 2r\cos\theta + r^2)^{-\lambda} = \sum_{j=0}^{\infty} r^j P_j^\lambda(\cos\theta)$, *for* $0 \leq r < 1$, *where the* P_j^λ *are ultraspherical polynomials. Then, for* $0 <$ Re $\lambda < 1/2$, *we have*

$$|(1 - 2r\cos\theta + r^2)^{-\lambda} - \sum_{j=0}^{m-1} r^j P_j^\lambda(\cos\theta)| \leq C_\lambda r^{m-1+2\,\mathrm{Re}\,\lambda} \cdot |(1 - 2r\cos\theta + r^2)|^{-\mathrm{Re}\,\lambda}$$

for $0 < r < \infty$, *with* C_λ *independent of* m.

To obtain the pointwise bound for $K_z(x,y)$ from Lemma 1.5, we let $r = |x|$, $w = \frac{x}{|x|}$, $s = |y|$, $w' = \frac{y}{|y|}$, $\cos\theta = w \cdot w'$, and then

$$|x-y|^{-n+z} = (r^2 + s^2 - 2rs\cos\theta)^{(z-n)/2} =$$

$$= s^{z-n} \left(1 + \left(\frac{r}{s}\right)^2 - 2\cos\theta \left(\frac{r}{s}\right)\right)^{(z-n)/2}.$$

Thus, by the formula for $K_z(x,y)$, given before (1.2), Lemma 1.5 gives the desired pointwise bound.

Lemma 1.5 can be proved as follows: Let $f(z) = \sum_{j=0}^{\infty} a_j z^j$ be an analytic function in $|z| < 1$. Then, $f(z) - \sum_{j=0}^{m-1} a_j z^j = \frac{z^m}{2\pi i} \int_{|\varsigma|=1} \frac{f(\varsigma)\varsigma^{-m}}{\varsigma - z} d\varsigma$. We then let $f(z) = \{(1 - ze^{i\theta})(1 - ze^{-i\theta})\}^{-\lambda}$, and show that $\sup_m \left|\int_{-\pi}^{\pi} f(e^{it}) e^{-imt} (e^{it} - r)^{-1} dt\right| \leq C_\lambda |1 - 2r \cos\theta + r^2|^{-\mathrm{Re}\,\lambda}$ for $0 < r < 2$, $0 <$ Re $\lambda < 1/2$. (See [52] for the details.)

Let me now return to the work of Sawyer ([41]), and Chanillo and Sawyer ([12]), mentioned in the Introduction. In [41], Sawyer proved Lemma 1.5 in the "end point" case Re $\lambda = 1/2$. From this, it follows that, if $\alpha = n - 1$,

$$\left| |x - y|^{-n+\alpha} - \sum_{j=0}^{m-1} a_{j,\alpha}(x, y) \right| \leq C|x|^m |y|^{-m} |x - y|^{-n+\alpha}.$$

When $n = 3$, $\alpha = 2$, $-n + \alpha = -1 = -(n - 2)$, which explains why the results are easier for $n = 3$. It is also easy to see that the above inequality is false for $\alpha < n - 1$. Also, this estimate easily shows that when $n = 3$, $\lim_{r \to 0} \sup_{x \in K} \int_{|x-y|<r} \frac{|V(y)|}{|x - y|} dy = 0$, $-\Delta + V$ has (s.u.c.p.).

Let us now go to [12]. Thus, we assume that

$$\lim_{r \to 0} \sup_{x \in K} \left(\frac{1}{|B_r|} \int_{B_r} |V|^p \right)^{1/p} r^2 = 0, \quad p > \frac{n - 1}{2}.$$

We assume that $|\Delta u| \leq |Vu|$, and that u vanishes of ∞ order at 0. We want to conclude that $u \equiv 0$. We can clearly assume $V \geq 0$.

Define now

$$S_z(f)(x) = \frac{V^{z/4}(x)}{|x|^N} \int_{\mathbb{R}^n} f(y)|y|^N V^{z/4}(y) R_z(x, y) dy$$

where

$$R_z(x, y) = \left[\frac{1}{|x - y|^{n-z}} - \sum_{j=0}^{N-1} \frac{1}{j!} \left(\frac{\partial}{\partial s} \right)^j |sx - y|^{-n+z}|_{s=0} \right].$$

When Re $z = 0$, one can use Lemma 1.3 to show that

$$\|S_z(f)\|_{L^2(\mathbb{R}^n)} \leq C\|f\|_{L^2(\mathbb{R}^n)}.$$

On the other hand, for Re $z = \alpha$, $n - 1 < \alpha < n$, Lemma 1.5 shows that

$$|S_z f(x)| \leq C_z V^{\alpha/4}(x) \int_{\mathbb{R}^n} |f(y)| \frac{V^{\alpha/4}(y) dy}{|x - y|^{n-\alpha}}.$$

Let I_α be the fractional integration operator, given by convolution with $|x - y|^{-n+\alpha}$. Then, when $n - 1 < \alpha < n$, we have

$$\int_{\mathbb{R}^n} |S_z f(x)|^2 dx \leq C_z \int_{\mathbb{R}^n} |I_\alpha(|f|V^{\alpha/4})|^2 V^{\alpha/2} dx.$$

We are thus led to understand inequalities of the form

$$\int I_\alpha(f)^2 w \leq C \int f^2 w^{-1}. \tag{1.6}$$

By the semigroup property of fractional integration (see [50], for instance), (1.6) is equivalent to

$$\int (I_{\alpha/2} g)^2 w \leq \sqrt{C} \int g^2,$$

and its dual

$$\int (I_{\alpha/2} f)^2 \leq \sqrt{C} \int f^2 w^{-1}. \tag{1.7}$$

(1.7) is called a trace inequality for $I_{\alpha/2}$, and has been extensively studied ([17],[33]). In particular, [33] gives necessary and sufficient conditions on w for (1.7) to hold, while C. Fefferman and Phong ([17]) show that if

$$\sup_{B_r} r^\alpha \left(\frac{1}{|B_r|} \int_{B_r} w^q \right)^{1/q} \leq C_q, \tag{1.8}$$

for some $q > 1$, and constant C_q, (1.7) holds with $\sqrt{C} \leq AC_q$. (The results in [13] also give sufficient conditions for (1.6) to hold, which can be applied to our case.) We then notice that $w(x) = V^{\alpha/2}(x)$ verifies (1.8) for some $q > 1$, for some $n-1 < \alpha < n$, because $p > \dfrac{n-1}{2}$ in our original assumption on V. Thus, by complex interpolation ([49]),

$$\int_{\mathbb{R}^n} |S_2 f|^2 dx \leq C \int_{\mathbb{R}^n} |f|^2 dx.$$

Note also that, if we replace V by $V_\rho = V\chi_{B(0,\rho)} + \dfrac{\varepsilon}{|x|^2}\chi_{{}^c B(0,\rho)}$, and we repeat the above argument, we obtain the same inequality, but with constant C which can be chosen as small as we please, by choosing ρ, and then ε sufficiently small. This follows from our "vanishing" assumption on V, and the remark after (1.8).

Now, let $\eta \in C_0^\infty(\mathbb{R}^n)$, $\eta \equiv 1$ on $B(0,\rho)$. Then, $\dfrac{\eta u V_\rho^{1/2}}{|x|^N} = CS_2(\Delta(\eta u)|y|^{-N} V_\rho^{-1/2})$. Thus,

$$\int_{B(0,\rho)} u^2 \frac{V}{|x|^{2N}} \leq C \int_{\mathbb{R}^n} |S_2(\Delta(\eta u)|y|^{-N} V_\rho^{-1/2})|^2 \leq$$

$$\leq C \int_{\mathbb{R}^n} |\Delta(\eta u)|y|^{-N}|^2 V_\rho^{-1} dy,$$

where C is small. Splitting the last integral into $B(0,\rho)$ and ${}^C B(0,\rho)$, and using the fact that in $B(0,\rho)$, $|\Delta u| \leq |Vu|$, and that C is small, we obtain

$$\int_{B(0,\rho)} u^2 \frac{V}{|x|^{2N}} dx \leq 2C \int_{{}^c B(0,\rho)} |\Delta(\eta u)|^2 |y|^{-2N} V_\rho^{-1}(y) dy.$$

Again, letting $N \to \infty$, we obtain $u \equiv 0$ in $B(0,\rho)$.

Section 2: Carleman estimates and discrete restriction theorems.

In this section we will sketch D. Jerison's ([27]) proof of (0.2), using the discrete restriction inequality (0.5) of C. Sogge ([46]).

Thus, let us return to the inequality (1.1), or its equivalent form (1.2). Recall, from the sketch of the proof of Lemma 1.3, that if $g \in C_0^\infty(\mathbb{R}_+)$, and $P_k(w)$ is a spherical harmonic of degree k,

$$T_2(g(r)P_k(w)) = T_{2,k}(g)P_k(w),$$

where

$$\begin{aligned} T_{2,k}(g)^\sim(\eta) &= C\frac{\Gamma\left(\frac{1}{2}(k-t-i\eta)\right)}{\Gamma\left(\frac{1}{2}(k-t-2-i\eta)\right)}\cdot\frac{\Gamma\left(\frac{1}{2}(n+k+t-2+i\eta)\right)}{\Gamma\left(\frac{1}{2}(n+k+t+i\eta)\right)}\cdot\tilde{g}(\eta) = \\ &= C\frac{1}{(k-t+i\eta)}\cdot\frac{1}{(k+t+(n-2)+i\eta)}\tilde{g}(\eta), \end{aligned}$$

by the functional equation $\Gamma(\varsigma+1) = \varsigma\Gamma(\varsigma)$.

It is now convenient to introduct logarithmic polar coordinates $x = e^y w$, $w \in S^{n-1}$, $y \in \mathbb{R}$, and to let

$$\hat{f}(\eta, w) = \int_{-\infty}^{+\infty} e^{i\eta y} f(y,w)\,dy.$$

Let now $\sigma_t(\eta,k) = \dfrac{1}{(k-t+i\eta)}\cdot\dfrac{1}{(k+t+(n-2)+i\eta)}$, and ξ_k be the projection operator from $L^2(S^{n-1})$ to the space of spherical harmonics of degree k. For $f \in C_0^\infty(\mathbb{R}\times S^{n-1})$, let

$$R_t f(y,w) = \sum_{k=0}^{\infty}\frac{1}{2\pi}\int_{-\infty}^{+\infty}\sigma_t(\eta,k)e^{-iy\eta}\xi_k\hat{f}(\eta,-)(w)\,d\eta. \tag{2.1}$$

It is then easy to see that (1.2) is equivalent to

$$\|R_t f\|_{L^{p'}(\mathbb{R}\times S^{n-1})} \le C\|f\|_{L^p(\mathbb{R}\times S^{n-1})}, \tag{2.2}$$

where $\dfrac{1}{p}+\dfrac{1}{p'} = 1$, $\dfrac{1}{p}-\dfrac{1}{p'} = \dfrac{2}{n}$.

Fix N so that $2^N \le t/10 \le 2^{N+1}$, and let $\{\theta_\beta\}_{\beta=0}^N$ be a smooth partition of unity of the positive real axis, with supp $\theta_0 \subset \{r : r \le 1\}$, supp $\theta_N \subset \{r : r > t/400\}$, and supp $\theta_\beta \subset \{r : 2^{\beta-2} \le r \le 2^\beta\}$, $0 < \beta < N$. For each β, consider the operator R_t^β, defined as in (2.1), but with symbol $\sigma_t^\beta(\eta,k) = \theta_\beta(|k-t+i\eta|)\sigma_t(\eta,k)$. Note that for $\beta \le N-1$, $\sigma_t^\beta(\eta,k)$ is supported where $|k-t+i\eta| \le 2^\beta$, and hence, there are at most $2^{\beta+1}$ terms in the sum over k which defines R_t^β. Moreover, the value of k is, in each case, comparable to t. It is also easy to check that

$$\left|\left(\frac{\partial}{\partial\eta}\right)^j \sigma_t^\beta(\eta,k)\right| \le C_j 2^{-(1+j)\beta} t^{-1}, \quad j = 0,1,\ldots. \tag{2.3}$$

Thus, if we use (2.3) and (0.5), we see that, for $\beta = 0, 1, \ldots, N-1$,

$$\left\| \sum_{k=0}^{\infty} \left(\frac{\partial}{\partial \eta}\right)^j \sigma_t^\beta(\eta, k)\xi_k g \right\|_{L^{p'}(S^{n-1})} \leq C_j 2^{-j\beta} t^{-2/n} \|g\|_{L^p(S^{n-1})}.$$

If we now write

$$R_t^\beta f(y, w) = \int_{-\infty}^{+\infty} K_t^\beta(y - y') f(y', -)(w) dy',$$

where

$$K_t^\beta(z) = \sum_{k=0}^{\infty} \frac{1}{2\pi} \int_{-\infty}^{+\infty} \sigma_t^\beta(\eta, k) e^{iz\eta} d\eta \xi_k =$$

$$= \sum_{k=0}^{\infty} \frac{1}{2\pi} \int_{-\infty}^{+\infty} \left(\frac{\partial}{\partial \eta}\right)^j \sigma_t^\beta(\eta, k) \frac{1}{(iz)^j} e^{iz\eta} d\eta \xi_k,$$

and we notice that, for $\beta \leq N-1$, the integration in η takes place over an interval of length smaller than 22^β, we see that, for each z, $K_t^\beta(z)$ is an operator from $L^p(S^{n-1})$ to $L^{p'}(S^{n-1})$ with norm less than or equal to $C_j|2^\beta z|^{-j} t^{-n/2} 2^\beta$. Using the values $j = 0$, and $j = 10$, we find that the norm is bounded by $C 2^\beta t^{-n/2}(1 + |2^\beta z|)^{-10}$.

Let now r verify $\frac{1}{r} + \frac{1}{p} = \frac{1}{p'} + 1$, i.e. $\frac{1}{r} = 1 - \frac{2}{n}$. Since

$$\|(1 + |2^\beta z|)^{-10}\|_{L^r(dz)} \leq C(2^\beta)^{-1/r},$$

Minkowski's integral inequality and Young's inequality give

$$\|R_t^\beta f\|_{L^{p'}(\mathbb{R} \times S^{n-1})} \leq C 2^\beta t^{-n/2} (2^\beta)^{2/n-1} \|f\|_{L^p(\mathbb{R} \times S^{n-1})}$$

for $\beta \leq N-1$. But, $\sum_{\beta=0}^{N-1} 2^{2\beta/n} \leq t^{n/2}$, and we have thus bounded $\sum_{\beta=0}^{N-1} R_t^\beta$. Finally, R_t^N can be easily controlled by ordinary fractional integration of order 2, and thus (2.2) follows.

Section 3: Weak unique continuation, uniform Sobolev inequalities and restriction theorems.

We start out by illustrating how the (w.u.c.p.) for $-\Delta u + Vu = 0$, with $V \in L^{n/2}(\mathbb{R}^n)$ can be applied to certain uniqueness questions in non-linear p.d.e. We are indebted to H. Brézis for stimulating conversations on this topic.

A well-known result of Pohozaev ([38], 1965) states that if Ω is a bounded starlike Lipschitz domain, and u is a non-negative weak solution of $-\Delta u = u^{n+2/n-2}$, which is 0 on $\partial\Omega$ in the weak sense (i.e. $u \in H_0^1(\Omega)$ = the closure of $C_0^\infty(\Omega)$ under the norm $\int_\Omega |\nabla u|^2$), then $u \equiv 0$. Here, the power $\frac{n+2}{n-2}$ is critical in the sense that there is lack of existence for $p \geq \frac{n+2}{n-2}$, and existence for $p < \frac{n+2}{n-2}$. Moreover, the assumption of starlikeness is essential ([6]). Pohozaev established his result be means of the following integral identity: If $-\Delta u = g(u)$ on Ω, u is 0 on $\partial\Omega$, and $G(u) = \int_0^u g(t)dt$, then

$$\left(1 - \frac{n}{2}\right) \int_\Omega g(u)u + n \int_\Omega G(u) = \frac{1}{2} \int_{\partial\Omega} \langle Q, N_Q \rangle \left(\frac{\partial u}{\partial N}\right)^2 d\sigma.$$

Here N_Q is the unit normal to $\partial\Omega$, and $\frac{\partial}{\partial N}$ denotes differentiation along the normal direction. The formula follows by applying Stokes' theorem to $\operatorname{div}((x \cdot \nabla u)\nabla u)$, and using the equation and boundary condition. To obtain Pohozaev's result, take $g(u) = u^{n+2/n-2}$, $G(u) = \frac{n-2}{2n} u^{2n/n-2}$, so that

$$\left(1 - \frac{n}{2}\right) \int_\Omega u^{n+2/n-2} u + \frac{n-2}{2} \int_\Omega u^{2n/n-2} = \frac{1}{2} \int_{\partial\Omega} \langle Q, N_Q \rangle \left(\frac{\partial u}{\partial N}\right)^2.$$

Thus, the last integral is 0, and by the starlike assumption $\frac{\partial u}{\partial N}|_{\partial\Omega} \equiv 0$. But then,

$$0 = -\int_\Omega \Delta u = \int_\Omega u^{n+2/n-2}, \text{ and hence, } u \equiv 0.$$

Let us now consider the case of variable sign solutions u. The appropriate equation is then $-\Delta u = |u|^{n+2/n-2} \operatorname{sign} u$. Proceeding as before, $G(u) = \frac{n-2}{2n}|u|^{2n/n-2}$ and hence, once again $\frac{\partial u}{\partial N}$ is 0 on $\partial\Omega$. But now, this only says that $0 = \int_\Omega |u|^{n+2/n-2} \operatorname{sign} u$. To show that $u \equiv 0$ in Ω, we extend u to be 0 outside of Ω. Since u, $\frac{\partial u}{\partial N}$ are 0 on $\partial\Omega$, the extended $u \in H_0^1(\mathbb{R}^n)$, and verifies the equation $-\Delta u = |u|^{n+2/n-2} \operatorname{sign} u$ in all of $\mathbb{R}^n$. Moreover, it has compact support. Let $V = |u|^{n+2/n-2-1}$. Then, since $u \in H_0^1(\mathbb{R}^n)$, the Sobolev embedding theorem shows that $V \in L^{n/2}(\mathbb{R}^n)$. Moreover, $-\Delta u = Vu$, and so by (w.u.c.p.) $u \equiv 0$.

Let us now turn to an alternative proof of the (w.u.c.p.). As we mentioned in the introduction, the (w.u.c.p.) for $-\Delta + V(x)$, $V(x) \in L^{n/2}(\mathbb{R}^n)$ is an easy consequence of the Carleman estimate

$$\|e^{-\tau x_n} u\|_{L^{p'}(\mathbb{R}^n)} \le C \|e^{-\tau x_n} \Delta u\|_{L^p(\mathbb{R}^n)}. \tag{3.1}$$

In fact, assume that $|\Delta u| \le |Vu|$, supp $u \subset \{x_n > 0\}$, and $V \in L^{n/2}(\mathbb{R}^n)$. Choose $\rho > 0$ so that, with C as in (3.1), $C\|V\|_{L^{n/2}(S_\rho)} \le 1/2$, where $S_\rho = \{x : 0 \le x_n \le \rho\}$. We will show that $u \equiv 0$ on S_ρ, and this will suffice. By (3.1), for $\tau > 0$

$$\begin{aligned} \|e^{-\tau x_n} u\|_{L^{p'}(S_\rho)} &\le C\|e^{-\tau x_n} \Delta u\|_{L^p(\mathbb{R}^n)} \le C\|e^{-\tau x_n} V u\|_{L^p(S_\rho)} + \\ &+ C\|e^{-\tau x_n} \Delta u\|_{L^p(\{x_n > 0\} \setminus S_\rho)} \le \\ &\le \frac{1}{2}\|e^{-\tau x_n} u\|_{L^{p'}(S_\rho)} + Ce^{-\tau\rho}\|\Delta u\|_{L^p(\{x_n > 0\})}, \end{aligned}$$

by Hölder's inequality. Thus,

$$\|e^{-\tau(x_n-\rho)}u\|_{L^{p'}(S_\rho)} \leq 2C\|\Delta u\|_{L^p(\{x_n>0\})}.$$

Letting $\tau \to +\infty$ we see that $u \equiv 0$ in S_ρ.

The substitution $v = e^{-\tau x_n}u$ shows that (3.1) is equivalent to the Sobolev inequality

$$\|v\|_{L^{p'}(\mathbb{R}^n)} \leq C\|[\Delta - 2\tau\frac{\partial}{\partial x_n} + \tau^2]v\|_{L^p(\mathbb{R}^n)}. \tag{3.2}$$

We are thus led to proving "uniform" Sobolev inequalities. We will now sketch the work of Kenig, Ruiz and Sogge ([31]). Let $P(D)$ be a complex coefficient, second order constant coefficient differential operator whose principal part $Q(D)$ is real and non-degenerate. We can assume, without loss of generality, that $Q(\xi) = -\xi_1^2 - \xi_2^2 - \quad -\xi_j^2 + \xi_{j+1}^2 + \quad + \xi_n^2$. We then have ([31])

$$\|v\|_{L^{p'}(\mathbb{R}^n)} \leq C\|P(D)v\|_{L^p(\mathbb{R}^n)}, n \geq 3, \frac{1}{p} - \frac{1}{p'} = \frac{2}{n}. \tag{3.3}$$

This inequality, then in turn, implies "global" unique continuation theorems for operators whose principal part is not necessarily elliptic. For instance, if p is as in (3.3), $\left(\frac{\partial^\alpha u}{\partial x^\alpha}\right)_{|\alpha|=2} \in L^p(\mathbb{R}^n)$, $|P(D)u| \leq |Vu|$, $V \in L^{n/2}(\mathbb{R}^n)$, and u vanishes on one side of a hyperplane, $u \equiv 0$. We call this theorem "global" because of the assumption $\left(\frac{\partial^\alpha u}{\partial x^\alpha}\right)_{|\alpha|=2} \in L^p(\mathbb{R}^n)$. Some global assumption is necessary when Q is not elliptic, because if the hyperplane is characteristic for $P(D)$ (i.e. $Q(N) = 0$, where N is the normal to the hyperplane), then there exists a $u \in C^\infty(\mathbb{R}^n)$, u vanishes on one side of the hyperplane, and $P(D)u = 0$ ([25]). However, our "global" assumptions are of course, not verified for u. In the case when $Q(D) = \frac{\partial^2}{\partial x_1^2} + \frac{\partial^2}{\partial x_2^2} + \quad + \frac{\partial^2}{\partial x_{n-1}^2} - \frac{\partial^2}{\partial x_n^2}$ is the wave operator (and hence the "spheres" $\tilde{S}^{n-1} = \{(x', x_n) \in \mathbb{R}^n : x_n^2 - |x'|^2 = 1, x_n > 0$ are strictly convex), one can use reflection along these spheres to obtain, from (3.3), a more local result:

Let $X \subset \mathbb{R}^n$, $n \geq 3$ be an open, proper, convex cone, with vertex y, such that no hyperplane through y which is characteristic with respect to $Q(D)$ intersects $\overline{X}$ only at y (for example, X can be a light cone). Suppose that $\left(\frac{\partial^\alpha u}{\partial x^\alpha}\right)_{|\alpha|=2} \in L^p_{\text{loc}}(X)$, with p as above, and that $|P(D)u| \leq |Vu|$, where $V \in L^{n/2}_{\text{loc}}(X)$. Then, if u vanishes outside a bounded subset of X, u must vanish in all of X. For the details, see [31].

There are also "parabolic" versions of (3.3). For example, Kenig and Sogge ([32]) have shown

$$\|u(x,t)\|_{L^{p'}(\mathbb{R}^{n+1})} \leq C\|[i\frac{\partial}{\partial t} + \Delta + \vec{a} \cdot \nabla_x + b]u(x,t)\|_{L^p(\mathbb{R}^{n+1})}. \tag{3.4}$$

Here $\frac{1}{p} - \frac{1}{p'} = \frac{2}{n+2}$, $\vec{a} \in \mathbb{C}^n$, $b \in \mathbb{C}$. A consequence of (3.4) is the following "global" unique continuation result for Schrödinger equations: Suppose that $\left|\left(i\frac{\partial u}{\partial t} + \Delta\right)u(x,t)\right| \leq |V(x,t)u(x,t)|$

in $\mathbb{R}^{n+1}$, that $\{((1+\tau^2)^{1/2}+|\xi|^2)\hat{u}(\xi,\tau)\}^\vee \in L^p(\mathbb{R}^{n+1})$, $V \in L^{n+2/2}(\mathbb{R}^{n+1})$, and that u vanishes in some half-space of $\mathbb{R}^{n+1}$. Then, $u \equiv 0$.

Let us now give a sketch of the proof of (3.3). The difficulty in proving (3.3) comes from the fact that $P(\xi)$, the symbol of $P(D)$, may vanish away from the origin. However, if this is the case, the zero set of $P(\xi)$ always lies on a "sphere", which explains the relevance of restriction theorems for the Fourier transform.

Let $H^n_\pm$ be the open subsets of $\mathbb{R}^n$ on which Q is strictly positive and negative respectively, and let $S^{n-1}_\pm$ be the "spheres" $S^{n-1}_\pm = \{\xi : Q(\xi) = \pm 1\}$. There are canonical measures $dw_\pm$ on $S^{n-1}_\pm$, so that, on $H^{n-1}_\pm$, $d\xi = \rho^{n-1}\,d\rho\,dw_\pm$.

The main tool in the proof of (3.3) is the following "restriction" lemma:

LEMMA 3.5. *Let $n \geq 3$, $\frac{1}{p} - \frac{1}{p'} = \frac{2}{n}$, Q as above. Then, for $f \in C_0^\infty(\mathbb{R}^n)$,*

(a) $\| \int_{S^{n-1}_\pm} \hat{f}(w) e^{ix\cdot w} dw_\pm \|_{L^{p'}(\mathbb{R}^n)} \leq C \|f\|_{L^p(\mathbb{R}^n)}$.

(b) *There exists a constant C, so that, for all $z \in \mathbb{C}$*

$$\|f\|_{L^{p'}(\mathbb{R}^n)} \leq C \|[Q(D)+z]f\|_{L^p(\mathbb{R}^n)}.$$

When $Q = -\Delta$, (a) is due to Stein–Tomas [56], while the general case is due to Strichartz [55]. As we shall see, (b) can be proven in the same way as (a). Moreover, it is not hard to see that choosing $z = \pm 1 + i\epsilon$, taking imaginary parts, and letting $\epsilon \to 0$, allows one to recover (a) from (b).

Let us assume Lemma 3.5 for the moment, and use it to prove (3.3). It is not hard to see that it is enough to prove (3.3) when $P(D) = Q(D) + \sigma + \epsilon\left\{\frac{\partial}{\partial x_j} + \beta\right\}$, $\sigma = \pm 1$, $\epsilon, \beta \in \mathbb{R}\backslash\{0\}$, $j = 1$ or n. We will deal with the case $\sigma = 1$, $j = n$, the other cases being similar. We are then reduced to proving the multiplier inequality

$$\left\| \left\{ \frac{\hat{f}(\xi)}{Q(\xi)+1+i\epsilon[\xi_n+\beta]} \right\}^\vee \right\|_{p'} \leq C\|f\|_p. \tag{3.6}$$

Let $m(\xi)$ be the multiplier in (3.6), and let $\chi(t)$ be 1 if $t \in [1,2]$, and 0 otherwise. Set $\chi_k(\xi_n) = \chi(2^k(\xi_n+\beta))$, and define $m_k(\xi) = \chi_k(\xi_n)m(\xi)$. It suffices to show

$$\|\{m_k(\xi)\hat{f}(\xi)\}^\vee\|_{p'} \leq C\|f\|_p. \tag{3.7}$$

In fact, noting that $p < 2 < p'$, Littlewood–Paley theory ([50]) and Minkowski's inequality, imply,

assuming (3.7), the following string of inequalities:

$$\|\{m(\xi)\hat{f}(\xi)\}^\vee\|_{p'} \le C\left\|\left(\sum_{k=-\infty}^{+\infty}|\{m_k(\xi)\hat{f}(\xi)\}^\vee|^2\right)^{1/2}\right\|_{p'} \le$$

$$\le C\left(\sum_{k=-\infty}^{+\infty}\|\{m_k(\xi)\hat{f}(\xi)\}^\vee\|_{p'}^2\right)^{1/2} \le$$

$$\le C\left(\sum_{k=-\infty}^{+\infty}\|\{\chi_k(\xi)\hat{f}(\xi)\}^\vee\|_p^2\right)^{1/2} \le C\|f\|_p.$$

In order to establish (3.7), we will use (3.5)(b), in the case when $z = 1 + i\varepsilon 2^{-k}$. Then,

$$\left\|\left\{\frac{\chi_k(\xi_n)}{Q(\xi)+1+i\varepsilon 2^{-k}}\hat{f}(\xi)\right\}^\vee\right\|_{p'} \le C\|f\|_p.$$

Thus, (3.7) will follow from the estimate

$$\left\|\left\{\frac{\chi_k(\xi_n)[i\varepsilon(\xi_n+\beta-2^{-k})]}{[Q(\xi)+1+i\varepsilon(\xi_n+\beta)][Q(\xi)+1+i\varepsilon 2^{-k}]}\hat{f}(\xi)\right\}^\vee\right\|_{p'} \le C\|f\|_p.$$

Let T_k be the multiplier operator in the above inequality, and let us use the polar coordinates $\xi = \rho w$ associated to Q. By Minkowski's integral inequality,

$$\|T_k f\|_{p'} \le \sum_{\pm}\int_0^\infty \rho^{n-1}\left\|\int_{S_\pm^{n-1}}\frac{\varepsilon\hat{f}(\rho w)\chi_k(\xi_n)(\xi_n+\beta-2^{-k})e^{i\rho w x}}{[\pm\rho^2+1+i\varepsilon(\xi_n+\beta)][\pm\rho^2+1+i\varepsilon 2^{-k}]}dw_\pm\right\|_{p'} d\rho.$$

Using (3.5)(a) and rescaling, we see that the right hand side is bounded by

$$C\int_0^\infty \rho^{n-1}\rho^{-2n/p'}\left\|\left\{\frac{\varepsilon\hat{f}(\xi)\chi_k(\xi_n)(\xi_n+\beta-2^{-k})}{[\pm\rho^2+1+i\varepsilon(\xi_n+\beta)][\pm\rho^2+1+i\varepsilon 2^{-k}]}\right\}^\vee\right\|_p d\rho.$$

Since the L^p multiplier norm of $\dfrac{\varepsilon\chi(2^k(\xi_n+\beta))(\xi_n+\beta-2^{-k})}{[\pm\rho^2+1+i\varepsilon(\xi_n+\beta)][\pm\rho^2+1+i\varepsilon 2^{-k}]}$ is easily seen to be bounded by $\dfrac{\varepsilon 2^{-k}}{(\rho^2-1)^2+(\varepsilon 2^{-k})^2}$, and $1 = n-1-2n/p'$, the right hand side is bounded by

$$C\int_0^\infty \frac{\rho\varepsilon 2^{-k}}{(\rho^2-1)^2+(\varepsilon 2^{-k})^2}d\rho\|f\|_p \le C\|f\|_p,$$

which establishes (3.7).

SKETCH OF THE PROOF OF (3.5)(b): Because $\dfrac{1}{p}-\dfrac{1}{p'}=\dfrac{2}{n}$, it suffices to show the inequality when $|z|\ge 1$. A limiting argument allows us to assume $\mathrm{Im}(z)\ne 0$. Thus, we have to show the multiplier inequality

$$\left\|\left\{\frac{\hat{f}(\xi)}{[Q(\xi)+z]}\right\}^\vee\right\|_{p'} \le C\|f\|_p, \frac{1}{p}-\frac{1}{p'}=\frac{2}{n}.$$

This inequality can be proven by complex interpolation. Thus, for $\lambda \in \mathbb{C}$, consider the analytic family of operators T_λ, given by the multipliers $m_\lambda(\xi) = [Q(\xi)+z]^\lambda$. T_{-1} is our multiplier, and by complex interpolation ([49]), the desired inequality will follow from the estimates

$$\|T_\lambda f\|_2 \leq C e^{C|\mathrm{Im}\ \lambda|^2}\|f\|_2, \text{ Re } \lambda = 0$$

$$\|T_\lambda f\|_\infty \leq C e^{C|\mathrm{Im}\ \lambda|^2}\|f\|_1, \text{ Re } \lambda = -n/2.$$

The first estimate is obvious, while the second one follows from the fact that, if F_λ is the Fourier of m_λ, then $\|F_\lambda\|_\infty \leq C e^{C|\mathrm{Im}\ \lambda|^2}$, Re $\lambda = -n/2$. To establish this last estimate, one can use the following formula for F_λ ([19]):

$$F_\lambda(x) = \frac{2^{\lambda+1} e^{-\pi i j/2}}{(2\pi)^{n/2}\Gamma(-\lambda)} \cdot \left[\frac{z}{Q(x)}\right]^{1/2[n/2+\lambda]} K_{n/2+\lambda}(\sqrt{zQ(x)}),$$

where j is the signature of Q, K_v is "Bessel function" which can be expressed, for $w \in \mathbb{C}$, Re $w > 0$ as

$$K_v(w) = \int_0^\infty e^{-w cht} chvt dt.$$

The relevant estimates for K_v are

$$|e^{v^2} v K_v(w)| \leq C|w|^{-|\mathrm{Re}\ v|}, |w| \leq 1, \text{ Re } w > 0$$

(let $u = e^t$ in the above definition), and

$$|K_v(w)| \leq C_{\mathrm{Re}\ v} e^{-\mathrm{Re}\ w}|w|^{-1/2}, |w| \geq 1, \text{ Re } w > 0.$$

This last estimate can be obtained from the following alternative formula for K_v ([19]), valid for Re $v \geq 0$:

$$\Gamma(v+1/2)K_v(w) = \left(\frac{\pi}{2w}\right)^{1/2} e^{-w} \int_0^\infty e^{-t} t^{v-1/2}\left(1+\frac{t}{2w}\right)^{v-1/2} dt.$$

This concludes the proof of (3.5).

In section 2, we saw how D. Jerison used C. Sogge's "discrete" restriction lemma for S^{n-1}, to derive Carleman inequalities in $\mathbb{R}^n$. In this section we saw how the Stein–Tomas restriction lemma in $\mathbb{R}^n$, and its generalizations by Strichartz also give Carleman estimates and uniform Sobolev inequalities on $\mathbb{R}^n$. Moreover, as we observed in the introduction, the uniform Sobolev inequalities in fact also imply the Stein–Tomas lemma. This point of view was taken up by C. Sogge ([47]), who showed how to use uniform Sobolev inequalities on manifolds, to extend his "discrete" restriction lemma to arbitrary compact manifolds. We will conclude this section with a brief description of Sogge's result.

Let M be a smooth, connected compact manifold, of dimension $n-1$, $n \geq 3$, let P be a second order elliptic operator on M, with smooth coefficients, self-adjoint with respect to a smooth density $d\mu$ on M, and with real principal part. Let $\{\lambda_j\}_{j=0}^{\infty}$ be the eigenvalues of P, increasingly ordered, and let ξ_k be the projection operators on $L^2(M, d\mu)$, associated to the sum of the eigenspaces $\sum_{j\in\Lambda_k} \mathcal{H}_j$, where $\Lambda_k = \{j : \sqrt{\lambda_j} \in [k-1, k)\}$, and $\mathcal{H}_j$ is the eigenspace associated with λ_j. Then, if $\frac{1}{p} - \frac{1}{p'} = \frac{2}{n}$, we have ([47])

$$\|\xi_k g\|_{L^{p'}(M)} \leq Ck^{1-2/n}\|g\|_{L^p(M)} \tag{3.8}$$

as in the case of S^{n-1} ((0.5)).

One first observes that (3.8) follows from

$$\|\xi_k g\|_{L^2(M)} \leq Ck^{1/2-1/n}\|g\|_{L^p(M)}. \tag{3.9}$$

In fact, since ξ_k is a projection operator, (3.9) is equivalent, by duality, to

$$\|\xi_k g\|_{L^{p'}(M)} \leq Ck^{1/2-1/n}\|g\|_{L^2(M)}, \tag{3.10}$$

and applying (3.9) once more, we obtain (3.8).

Finally, (3.10) is in turn, implied by the "uniform" Sobolev inequality

$$\|u\|_{L^{p'}(M)} \leq Ck^{1/2-1/n-1}\|(P-k^2)u\|_{L^2(M)} + Ck^{1/2-1/n}\|u\|_{L^2(M)}. \tag{3.11}$$

To see that (3.10) follows from (3.11), apply (3.11) to $u = \xi_k g$. Then, by our definition of ξ_k, it is easy to see that

$$\|(P-k^2)\xi_k g\|_{L^2(M)} \leq Ck\|g\|_{L^2(M)}.$$

The advantage of (3.11) over (3.10) is that it can be proven locally, by using appropriate coordinates (Hadamard coordinates) ([25], Vol III, 17.4) and results of Carleson–Sjölin and Stein ([51]) on oscillatory integrals on $\mathbb{R}^n$. As was mentioned in the introduction, C. Sogge ([48]) used (3.8) as one of the main tools to extend results on Bochner–Riesz summability in $\mathbb{R}^n$, to arbitrary compact manifolds.

Section 4: Some open problems.

I would like to conclude this article by mentioning some open problems in this area.

(4.1) What is the optimal value of p, so that if $V \in J_p^{\text{loc}}$, (see the Introduction for the definition), $-\Delta + V(x)$ has the (s.u.c.p.). As was explained in Section 1, Chanillo and Sawyer ([12]) have shown this to be the case for $p > n - 1/2$.

(4.2) (Simon's conjecture ([**45**])). Does $V \in K_n^{\text{loc}}$ (again see the Introduction for the definition) imply that $-\Delta + V(x)$ has (s.u.c.p.), (u.c.p.) or (w.u.c.p.)? This was shown to be true by Sawyer ([**41**]) when $n = 3$.

(4.3) Variable coefficient problems: For instance, does (s.u.c.p.), or even (w.u.c.p.) hold for operators of the form $\sum a_{jk}(x)\dfrac{\partial^2}{\partial x_j \partial x_k} + V(x)$, where $\{a_{jk}(x)\}$ is an elliptic, symmetric, real, Lipschitz continuous (or even C^∞) matrix, and $V \in L_{\text{loc}}^{n/2}(\mathbb{R}^n)$? Hörmander ([**26**]) has established (u.c.p.) when $V \in L_{\text{loc}}^{4n-2/7}(\mathbb{R}^n)$. Another interesting question is whether there are any uniform Sobolev inequalities for $\sum a_{jk}(x)\dfrac{\partial^2}{\partial x_j \partial x_k}$.

(4.4) Gradient problems: For instance, is there (s.u.c.p.) for $\Delta + \overrightarrow{V}(x) \cdot \nabla$, $\overrightarrow{V} \in L_{\text{loc}}^n(\mathbb{R}^n)$? Hörmander ([**26**]) showed (u.c.p.) for $L_{\text{loc}}^p(\mathbb{R}^n)$, $p > \dfrac{3n-2}{2}$, while Barceló, Kenig, Ruiz and Sogge ([**7**]) showed (u.c.p.) for $p = \dfrac{3n-2}{2}$. Similar results were obtained earlier by D. Jerison ([**27**]) for the Dirac operator. The problem that one encounters is that

$$\||x|^{-\tau}\nabla u\|_{L^{p'}(\mathbb{R}^n)} \le C\||x|^{-\tau}\Delta u\|_{L^p(\mathbb{R}^n)},$$

$\dfrac{1}{p} - \dfrac{1}{p'} = \dfrac{1}{n}$, $n \ge 3$ is false. Moreover, $\|e^{-\tau\varphi}\nabla u\|_{L^q(U)} \le C\|e^{-\tau\varphi}\Delta u\|_{L^r(U)}$, U an open set in $\mathbb{R}^n$, φ non constant, $u \in C_0^\infty(U)$, can only hold if $\dfrac{1}{r} - \dfrac{1}{q} \le \dfrac{3n-2}{2}$ (see Jerison [**27**]).

(4.5) "Local" unique continuation for the heat equation: Suppose that $\left|\dfrac{\partial u}{\partial t} - \Delta u\right| \le |V(x,t)u(x,t)|$, where $V \in L_{\text{loc}}^{n+2/2}(\mathbb{R}^{n+1})$, and u vanishes in an open set $U \subset \mathbb{R}^{n+1}$. Does u vanish in the "horizontal component" of U, i.e. $\{(x,t) : (x_0,t) \in U \text{ for some } x_0\}$? Some results in this direction are due to Saut and Scheurer [**40**].

(4.6) Distinction between (w.u.c.p.), (u.c.p.) and (s.u.c.p.): We have seen examples of potentials V in $L_{\text{loc}}^p(\mathbb{R}^n)$, $p < n/2$ for which (s.u.c.p.) for $\Delta + V(x)$ fails. However, as far as we know, there are no known examples of potentials $V \in L_{\text{loc}}^1(\mathbb{R}^n)$ so that (u.c.p.) for $\Delta + V(x)$ fails. Does (w.u.c.p.) or (u.c.p.) hold whenever $V \in L_{\text{loc}}^1(\mathbb{R}^n)$?

REFERENCES

1. Agmon, S., "Unicité et convexité dans les problèmes differentials," Séminaire de Mathématiques Supérieures, No. 13, Les Presses de l'Université de Montréal, 1966.
2. ———, *Lower bounds for solutions of Schrödinger equations*, J. Analyse Math. **23** (1970), 1–25.
3. Amrein, W., Bérthier, A. and Georgescu, V., *L^p inequalities for the Laplacian and unique continuation*, Ann. Inst. Fourier (Grenoble) **31** (1981), 153–168.
4. Aronszajn, N., *A unique continuation theorem for solutions of elliptic partial differential equations or inequalities of second order*, J. Math. Pures Appl. **36** (1957), 235–249.
5. Aronszajn, N., Krzywcki, A. and Szarski, J., *A unique continuation theorem for exterior differential forms on Riemannian manifolds*, Ark. for Met. **4** (1962), 417–453.
6. Bahri, A. and J.M. Coron, *Sur une equation elliptique nonlinéaire avec l'exposant critique de Sobolev*, C.R. Acad. Sc. Paris **301** (1985), 345–348.

7. Barceló, B., Kenig, C., Ruiz, A. and Sogge, C., *Weighted Sobolev inequalities and unique continuation for the Laplacian plus lower order terms*, to appear, Ill. J. of Math.
8. Bérthier, A., *On the point spectrum of Schrödinger operators*, Ann. Sci. Ecole Norm. Sup., 4*e s* **15** (1982), 1–15.
9. Bérthier, A. and Georgescu, V., *Sur le proprieté de prolongement unique pour l'operateur de Dirac*, C.R. Acad. Sc. Paris, Ser A **291** (1980), 603–606.
10. Calderón, A.P., *Uniqueness in the Cauchy problem for partial differential equations*, Am. J. of Math. **80** (1958), 16–36.
11. Carleman, T., *Sur un problème d'unicité pour les systemes d'equations aux derivées partielles à deux variables indépendentes*, Ark. for Mat. **26B** (1939), 1–9.
12. Chanillo, S. and Sawyer, E., to appear.
13. Chanillo, S. and Wheeden, R., *L^p estimates for fractional integrals and Sobolev inequalities with applications to Schrödinger operators*, Comm. in Partial Diff. Eqtns. **10(9)** (1985), 1077–1116.
14. Cordes, H.O., *Über die Bestnuntheit der Lösunger elliptischer Differenticlgleichungen durch Anfangsvorgaben*, Nachr Akad. Wiss. Göttingen Math. Phys. K1, IIa **11** (1956), 239–258.
15. Erdelyi, A. (Director), "Higher Transcendental Functions," Bateman manuscript project, McGraw-Hill, New York, 1955.
16. Fefferman, C., *A note on spherical summation multipliers*, Israel J. Math. **15** (1973), 44–52.
17. ________, *The uncertainty principle*, Bull. A.M.S. **9(2)** (1983), 129–206.
18. Gorofalo, N. and Lin, F.H., *Monotonicity properties of variational integrals, A_p weights and unique continuation*, Indiana U. Math. J. **35** (1986), 245–268.
19. Gelfand, I.M. and Shilov, G.E., "Generalized Functions," Vol 1, Academic Press, New York, 1964.
20. Georgescu, V., *On the unique continuation property for Schrödinger Hamiltonians*, Helv. Phys. Acta. **52** (1979), 655–670.
21. Hadamard, J., "Le problème de Cauchy et les equations aux derivées partielles linéaires hyperboliques," Herman et Cie, Paris, 1932.
22. Hörmander, L., *On the uniqueness of the Cauchy problem*, Math. Scand. **6** (1958), 213–225.
23. ________, *On the uniqueness of the Cauchy problem II*, Math. Scand. **7** (1959), 177–190.
24. ________, *The spectral function of an elliptic operator*, Acta Math. **88** (1968), 341–370.
25. ________, "The Analysis of Linear Partial Differential Operators," Vol I, III, Springer-Verlag, New York, Berlin, 1983, 1985.
26. ________, *Uniqueness theorems for second order elliptic operators*, Comm. Par. Diff. Eqtns. **8** (1983), 21–64.
27. Jerison, D., *Carleman inequalities for the Dirac and Laplace operators, and unique continuation*, Advances in Math. **63** (1986), 118–134.
28. Jerison, D. and Kenig, C., *Unique continuation and absence of positive eigenvalues for Schrödinger operators*, Ann. of Math. **121** (1985), 463–494.
29. Kato, T., *Growth properties of solutions of the reduced wave equation with variable coefficients*, Comm. Pure and Appl. Math. **12** (1959), 403–425.
30. Kenig, C., *Carleman estimates, uniform Sobolev inequalities for second order differential operators, and unique continuation theorems*, to appear, Proc. of the Int. Congress of Math., Berkeley, California, 1986.
31. Kenig, C., Ruiz, A. and Sogge, C., *Uniform Sobolev inequalities and unique continuation for second order constant coefficient differential operators*, Duke Math. J. **55** (1987), 329–347.
32. Kenig, C. and Sogge, C., *A note on unique continuation for Schrödinger's operator*, to appear, Proc. A.M.S.
33. Kerman, R. and Sawyer, E., *Weighted norm inequalities for potentials with applications to Schrödinger operators, the Fourier transform and Carleson measures*, Bull. A.M.S. **12(1)** (1985), 112–116.
34. Muckenhoupt, B., *The equivalence of two conditions for weight functions*, Studia Math. **49** (1974), 101–106.
35. Müller, C., *On the behavior of the solution of the differential equation $\Delta u = f(x,u)$ in the neighborhood of a point*, Comm. Pure and Appl. Math. **1** (1954), 505–515.
36. Nirenberg, L., *Uniqueness in Cauchy problems for differential equations with constant leading coefficients*, Comm. Pure and Appl. Math. **10** (1957), 89–106.
37. Pliš, A., *On non-uniqueness in the Cauchy problem for an elliptic second order differential operator*, Bull. Acad. Pol. Sci. **11** (1963), 65–100.
38. Pohozaev, S.I., *Eigenfunctions of the equation $\Delta u + \lambda f(u) = 0$*, Soviet Math. Dokl. **5** (1965), 1408–1411.
39. Saut, J. and Scheurer, B., *Un théorème de prolongement unique pour des opérateurs elliptiques dont les coefficients ne sont pas localement bornés*, C.R. Acad. Sc. Paris, Ser A **290** (1980), 598–599.
40. ________, *Unique continuation for some evolution equations*, J. of Diff. Eqtns. **66** (1987), 118–139.
41. Sawyer, E., *Unique continuation for Schrödinger operators in dimension three or less*, Annales de l'Institut Fourier (Grenoble) **33** (1984), 189–200.
42. Schechter, M. and Simon, B., *Unique continuation for Schrödinger operators with unbounded potential*, J. Math. Anal. Apl. **77** (1980), 482–492.
43. Senator, K., *Unique continuation for Schrödinger equations in dimensions three and four*, Studia Math. **81** (1985), 311–321.

44. Simon, B., *On positive eigenvalues of one body Schrödinger operators*, Comm. Pure and Appl. Math. **22** (1969), 531–538.
45. ________, *Schrödinger semigroups*, Bull. A.M.S. **7(3)** (1982), 447–526.
46. Sogge, C., *Oscillatory integrals and spherical harmonics*, Duke Math. J. **53** (1986), 43–65.
47. ________, *Concerning the L^p norm of spectral clusters for second order elliptic operators on compact manifolds*, to appear, J. of Funct. Anal.
48. ________, *On the convergence of Riesz means on compact manifolds*, Ann. of Math. **126** (1987), 439–447.
49. Stein, E., *Interpolation of linear operators*, Trans. Amer. Math. Soc. **83** (1956), 482–492.
50. ________, "Singular integrals and differentiability properties of functions," Princeton U. Press, Princeton, New Jersey, 1972.
51. ________, "Oscillatory integrals in Fourier Analysis," in "Beijing Lectures in Harnonic Analysis", Annals of Math. Studies **112**, Princeton U. Press, Princeton, New Jersey, 1986, 307–356.
52. ________, *Appendix to "Unique continuation"*, Ann. of Math. **121** (1985), 489–494.
53. Stein, E. and Weiss, G., *Fractional integrals on n-dimensional Euclidean space*, J. of Math. and Mech. **7** (1958), 503–514.
54. ________________, "Introduction to Fourier Analysis on Euclidean Space," Princeton U. Press, Princeton, New Jersey, 1971.
55. Strichartz, R., *Restriction of Fourier transforms to quadratic surfaces*, Duke Math. J. **44** (1977), 705–714.
56. Tomas, P., *A restriction theorem for the Fourier transform*, Bull. A.M.S. **81** (1975), 477–478.
57. Von Neumann, J. and Wigner, E., *Über merkürdige diskrete Eigenwerte*, Phys. Z. **30** (1929), 465–467.

PROBLEMS IN HARMONIC ANALYSIS RELATED TO CURVES AND SURFACES WITH INFINITELY FLAT POINTS

Stephen Wainger

Section 0. Introduction.

A great deal of progress has been made on a variety of problems related to curves and surfaces. The most complete results on these problems are obtained when some type of curvature is assumed not to vanish to infinite order. In general, we might refer to this case as the non–flat case. If the curvature vanishes to infinite order we shall say we are in the flat case.

We shall consider here the following five classes of operators:

I. Hilbert transforms and maximal functions along curves: These are the operators defined for f in $C_0^\infty(R^n)$ by

(1) $$Hf(x) = \int_{-1}^{1} f(x-\gamma(t)) \frac{dt}{t}$$

and

(2) $$\mathcal{M}f(x) = \sup_{0<\epsilon\leq 1} \frac{1}{\epsilon}\left|\int_0^\epsilon f(x-\gamma(t))\frac{dt}{t}\right|$$

respectively, where $\gamma(t)$ is a curve in R^n with $\gamma(0) = 0$. Generally, we shall assume $\gamma(t)$ is smooth. We shall be concerned with the existence of L^p bounds for H and $\mathcal{M}$. The main results here in the non–flat case are due to Nagel, Riviere, Stein, and Wainger in the 1970's. (See [SW].) There are also some recent refinements due to Christ [C] and Christ and Stein [CS].

II. Let ϕ be a smooth homeomorphism of the circle onto the circle. For functions f on the circle we define

(3) $$T_\phi(f) = f(\phi(x)).$$

Supported in part by an N.S.F. grant at the University of Wisconsin.

The problem here is to determine those ϕ such that $T_\phi(f)$ has a uniformly convergent Fourier Series whenever f has an absolutely convergent Fourier Series. The results in the non–flat case are due to Alpar [A], Halasz [H] and Kaufman [KA]. See [K] for a discussion of this problem.

III. The Fourier Transform of compactly supported measures on convex hypersurfaces, S, in R^n. The problem here is to determine the decay rate of the Fourier Transform in terms of the geometry of S. We have in mind the results of Bruna, Nagel and Wainger [BNW] and Cowling and Mauceri [CM] in the non–flat case.

IV. Maximal operators on compact convex hypersurfaces in R^n: Let S be a smooth compact strictly convex hypersurface in R^n with surface measure $d\sigma$. Let $B(x,\epsilon)$ denote the set of points y in S at a distance at most ϵ from the tangent plane to S at x. We then consider

$$Mf(x) = \sup_{\epsilon>0} \frac{1}{|B(x,\epsilon)|} \int_{B(x,\epsilon)} |f(y)|\,dy \tag{4}$$

The problem is to show the mapping $f \to Mf$ is of weak type 1–1. In the non–flat case this is an immediate consequence of [BNW].

V. Restriction theorems for the Fourier Transform: E.M. Stein showed that for certain hypersurfaces, S, in R^n, the map R: $f \to \hat{f}|_S$ (where $g|_S$ means the restriction of g to S) is well defined on $L^p(R^n)$ for certain $p>1$. In the non–flat case, a great deal of literature was devoted to this phenomenon. Among the contributors were Fefferman and Stein [F], Sjölin [SJ], Christ [C1] and [C2], Drury [D] and Drury and Marshall [DM], Prestini [P], Ruiz [R], Strichartz [STR1], and Tomas [T].

We are concerned here with extensions to the case that S has a flat point.

In the flat case the most studied of these problems is I, and that will contribute the largest section of this paper. Sections 1–5 deal with a survey of progress on problems 1 to 5 above. In section 6 we try to explain the principal difficulties that arise in the flat case.

Section 1. Hilbert Transforms and maximal functions along curves.

For the most part we will consider the case of curves in the plane. We shall write

$\gamma(t) = (t,\Gamma(t))$, $\gamma(t)$ in R^2. Our operators (1) and (2) then become

(5) $$H_\Gamma f(x,y) = \int_{-1}^{1} f(x-t,\, y-\Gamma(t))\,\frac{dt}{t},$$

and

(6) $$\mathcal{M}_\Gamma f(x,y) = \sup_{\epsilon>0} \frac{1}{\epsilon}\,\Big|\int_0^\epsilon f(x-t,\, y-\Gamma(t)dt\Big|.$$

We shall assume throughout that $\Gamma(0) = \Gamma'(0) = 0$.

In studying $\mathcal{M}_\Gamma$, it is of course only necessary to consider $\Gamma(t)$ for $t\geq 0$. In the case of H_Γ, it is necessary to consider $\Gamma(t)$ for $t>0$ and $t<0$, and we will assume $\Gamma(t)$ is either even or odd. Finally, most of our discussion will focus on Γ's which are increasing and convex for $t>0$.

Let us examine the significance of the assumption that $\Gamma(t)$ is convex for positive t. First, for each vector (ξ,η) in R^2, let

$$E(\xi,\eta) = \{t\epsilon[-1,1] \mid \text{normal to } \gamma \text{ at } (t,\Gamma(t)) \text{ is in the direction } (\xi,\eta)\}.$$

Matters are greatly simplified when the sets $E(\xi,\eta)$ are simple. This is of course insured by convexity.

A second more concrete reason to assume $\Gamma(t)$ is convex is that the assumption insures boundedness properties of certain one-dimensional operators that are naturally associated to H_Γ and $\mathcal{M}_\Gamma$.

Let us assume that $f(x,y) = \phi(x)g(y)$ where $\phi(x)$ is in $C_0^\infty(R)$ and $\phi(x)$ is one on a large neighborhood of the origin. Then one can see that if $\mathcal{M}_\Gamma$ is bounded on $L^p(R^2)$,

$$\mathfrak{M}_\Gamma(g)(y) = \sup_{0<\epsilon\leq 1} \frac{1}{\epsilon}\int_0^\epsilon |g(y-\Gamma(t)|dt$$

is bounded on $L^p(R)$. So we might begin by imposing on Γ a condition that at least insures the $L^p(R)$ boundedness of $\mathfrak{M}_\Gamma$. Notice that by changing variable we obtain

$$\mathfrak{M}_\Gamma(g)(y) = \sup_{0<\epsilon\leq 1} \frac{1}{\epsilon}\int_0^{\Gamma(\epsilon)} g(y-u)\,\frac{du}{|\Gamma'(\Gamma^{-1}(u))|}.$$

Hence if $\Gamma(t)$ is increasing and convex $|\Gamma'(\Gamma^{-1}(u))|^{-1}$ is positive and decreasing. It then follows by a standard argument that $\mathfrak{M}_\Gamma(g)(y)$ is dominated by a constant multiple of the

Hardy Littlewood maximal function of g at y. So the convexity of Γ insures that $\mathfrak{M}_\Gamma$ is of weak type 1–1 and is bounded on $L^p(R)$.

Similarly, if H_Γ is bounded on $L^p(R^2)$, the transformation $\tilde{H}_\Gamma$, defined by

$$\tilde{H}_\Gamma g(y) = \int_{-1}^{1} g(y-\Gamma(t))\,\frac{dt}{t}$$

would be bounded on $L^p(R)$. If we take the Fourier transform and interchange the order of integration, we see that

$$(\tilde{H}_\Gamma g)^\wedge(\xi) = \tilde{m}(\xi)\hat{g}(\xi)$$

where

$$\tilde{m}(\xi) = \int_{-1}^{1} e^{i\xi\Gamma(t)}\,\frac{dt}{t}.$$

First, we shall see that if $\Gamma(t)$ is increasing for positive t (and even or odd) $\tilde{m}(\xi)$ is bounded. Note:

(7) $$\Gamma(t) = \int_0^t \Gamma'(s)ds \leq t\Gamma'(t).$$

So

$$\left|\int_{|t|\leq\Gamma^{-1}(\frac{1}{|\xi|})} e^{i\xi\Gamma(t)}\,\frac{dt}{t}\right| \leq \int_{|t|\leq\Gamma^{-1}(\frac{1}{|\xi|})} |e^{i\xi\Gamma(t)}-1|\,\frac{dt}{t}$$

$$\leq |\xi| \int_{|t|\leq\Gamma^{-1}(\frac{1}{|\xi|})} |\Gamma(t)|\,\frac{dt}{t} \leq |\xi| \int_{|t|\leq\Gamma^{-1}(\frac{1}{|\xi|})} |\Gamma'(t)|dt$$

$$\leq |\xi|\Gamma(\Gamma^{-1}(\tfrac{1}{|\xi|})) \leq 1.$$

Also, integration by parts shows

$$\left|\int_{t=\Gamma^{-1}(\frac{1}{|\xi|})}^{1} e^{i\xi\Gamma(t)}\,\frac{dt}{t}\right| \leq$$

$$\leq \frac{2}{|\xi|\Gamma'(\Gamma^{-1}(\frac{1}{|\xi|}))\Gamma^{-1}(\frac{1}{|\xi|})} + \frac{1}{|\xi|}\int_{\Gamma^{-1}(\frac{1}{|\xi|})}^{1} \frac{\Gamma''(t)}{[\Gamma'(t)]^2}\,\frac{dt}{t}$$

$$+ \frac{1}{|\xi|}\int_{\Gamma^{-1}(\frac{1}{|\xi|})}^{1} \frac{1}{\Gamma'(t)}\,\frac{dt}{t^2}$$

$$\leq \frac{4}{|\xi|\Gamma'(\Gamma^{-1}(\frac{1}{|\xi|}))\Gamma^{-1}(\frac{1}{|\xi|})} \leq 4$$

by (7).

Similarly,

$$\left| \int_{-1\leq t\leq -\Gamma^{-1}(\frac{1}{|\xi|})} e^{i\xi\Gamma(t)}dt \right| \leq 4,$$

so $\tilde{m}(\xi)$ is bounded. This of course proves that if Γ is convex $\tilde{H}_\Gamma$ is bounded on $L^2(R)$.

It is also not difficult to show that if Γ is convex, $\tilde{H}_\Gamma$ is bounded on $L^p(R)$ for $1<p<\infty$. We define $\tilde{H}_z f$ by the formula

$$(\tilde{H}_z f)^\wedge(\xi) = m_z(\xi)\hat{f}(\xi)$$

where

$$m_z(\xi) = \int (1+\xi^2\Gamma^2(t))^{z/2} e^{i\xi\Gamma(t)} \frac{dt}{t}.$$

Then an argument similar to the argument proving m bounded shows

$$|m_z(\xi)| \leq C(1+|z|), \text{ if } \operatorname{Re} z < \tfrac{1}{4}.$$

Moreover, one can show that if Re $z < 0$, $m_z(\xi)$ satisfies the hypothesis of the Marcinkiewicz multiplier theorem. One then concludes the L^p boundedness of $\tilde{H}$ by Stein's interpolation theorem.

It turns out that the L^p boundedness criteria are different for H_Γ and $\mathcal{M}_\Gamma$. We can gain some insight into the difference between H_Γ and $\mathcal{M}_\Gamma$ by considering related operators.

(8) $$\tilde{H}_\Gamma f(x,y) = \int_{-\infty}^{\infty} f(x-t, y-\Gamma(t)) \frac{dt}{t},$$

(where we assume for simplicity that $\Gamma(t)$ is odd), and

(9) $$\tilde{\mathfrak{M}}_\Gamma f(x,y) = \sup_{0<\epsilon<\infty} \frac{1}{\epsilon}\int_0^\epsilon |f(x-t,y-\Gamma(t)|dt.$$

Here, of course the integration involves the whole line, rather than just an interval. Now suppooc

$$\Gamma(t) = \begin{cases} 0, & 0\leq t\leq 1 \\ t-1, & 1\leq t \end{cases}$$ Then

$$\tilde{\mathfrak{M}}_\Gamma f(x,y) \leq M_1 f(x,y) + M_2 f(x-1,y),$$

where

$$M_1f(x,y) = \sup_{\epsilon>0} \frac{1}{\epsilon}\int_0^{\epsilon} |f(x-t,y)|\,dt$$

and

$$M_2f(x,y) = \sup_{\epsilon>0} \frac{1}{\epsilon}\int_0^{\epsilon} |f(x-t,y-t)|\,dt.$$

M_1f and M_2f are seen to be bounded on all L^p, $p>1$, by applying the one–dimensional Hardy Littlewood maximal function theory. So

$$\|\tilde{\mathfrak{M}}_\Gamma f\|_{L^p} \le C_p\|f\|_{L^p},$$

$1<p\le\infty$.

We can see on the other hand that $\tilde{H}_\Gamma$ is unbounded even in L^2 either analytically or geometrically. First of all,

$$(\tilde{H}_\Gamma f)^\wedge(\xi,\eta) = \tilde{m}(\xi,\eta)\hat{f}(\xi,\eta),$$

where

$$\tilde{m}(\xi,\eta) = \int_{-\infty}^{\infty} e^{i\xi t+i\eta\Gamma(t)}\,\frac{dt}{t}.$$

Now

$$\tilde{m}(\xi,\eta) = \text{Bounded} + \int_1^{\infty} \sin(\xi t+\eta\Gamma(t))\,\frac{dt}{t}$$

$$= \text{Bounded} + \int_1^{\infty} \sin(\xi t+\eta t-\eta)\,\frac{dt}{t}$$

$$= \text{Bounded} - \sin\eta\int_1^{\infty} \cos(\xi t+\eta t)\,\frac{dt}{t}.$$

The last integral does not remain bounded as $\xi+\eta\to 0$ with η bounded away from 0. So $\tilde{H}_\Gamma$ is unbounded on L^2.

To see the unboundedness of $\tilde{H}_\Gamma$ geometrically, consider applying $\tilde{H}_\Gamma$ to the characteristic function of a parallelogram, E_N, bounded by the x–axis, the line $y=-N$, the line $y=x$, and the line $y=x+1$.

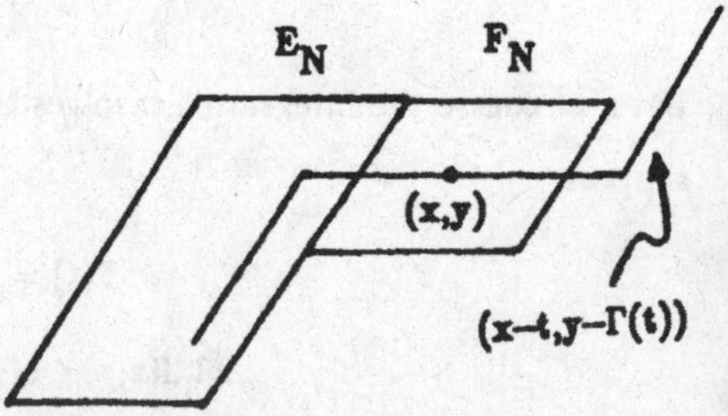

Note that in the parallelogram F_N, bounded by $y=0$, $y=\frac{-N}{2}$, $y=x$, and $y=x-1$. there is no cancellation between the part of the integral for $t>0$ with the part of the integral for $t<0$. Thus for (x,y) in F_N,

$$\tilde{H}f(x,y) \geq \log\frac{N}{2}.$$

Since F_N and E_N have comparable areas, $\tilde{H}$ cannot be bounded on any L^p space.

It turns out that an elaboration of these simple ideas allows one to construct examples of convex curves for which $\mathcal{M}_\Gamma$ is bounded and H_Γ is unbounded. As a matter of fact, if $\Gamma(t)$ for $t>0$ consists of a broken line so that

$$\Gamma(t) = m_j t + b_j, \qquad \alpha_{j+1} \leq t \leq \alpha_j,$$

H_Γ will be unbounded if the ratios α_j/α_{j+1} are unbounded, while $\mathcal{M}_\Gamma$ will be bounded on L^2 if $m_j \geq 2m_{j+1}$. If also, $\alpha_j \geq 2\alpha_{j+1}$, $\mathcal{M}_\Gamma$ is bounded on all L^p, $1<p\leq\infty$. See [NVWW1] and [CW].

There is another discrepancy besides the one between the maximal function and Hilbert transform. Suppose $\Gamma(t)$ is a given increasing convex curve for $t>0$. Let Γ^e denote the even extension of Γ giving rise to an operator

$$H_{\Gamma^e} = H_\Gamma^e.$$

Suppose also Γ is extended to an odd curve Γ^o, giving rise to an operator

$$H_{\Gamma^o} = H_\Gamma^o.$$

Then it is possible for H_Γ^o to be bounded on L^2 without H_Γ^e being bounded on L^2. Let us try to see very naively what the difference might be. Consider again applying the operator to the characteristic function of a tilted parallelogram.

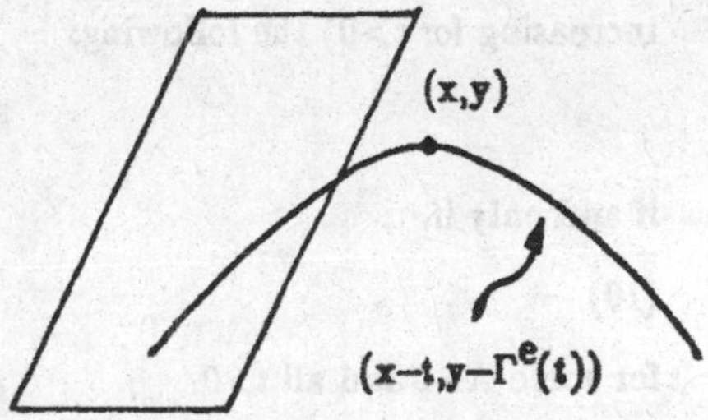

Notice that for (x,y) to the right of the parallelogram there is no chance for the integration with $t<0$ to cancel the integration with $t>0$. On the other hand, for the curve Γ^o, the picture is as follows:

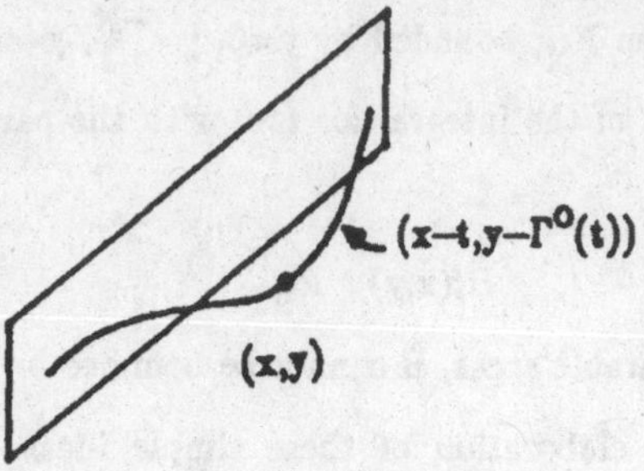

So conceivably, some cancellation can occur. Actually, the situation is much more complicated – but we have seen the basic reason, namely, in the case of odd curves, we can have more cancellation in the integral. Again, one can understand the situation analytically.

$$(H^o_\Gamma f)^\wedge(\xi,\eta) = \hat{f}(\xi,\eta) m^o_\Gamma(\xi,\eta)$$

and

$$(H^e_\Gamma f)^\wedge(\xi,\eta) = \hat{f}(\xi,\eta) m^e_\Gamma(\xi,\eta),$$

where

$$m_\Gamma^{\binom{o}{e}}(\xi,\eta) = \int_{-1}^{1} e^{i\xi t + i\eta\Gamma^{\binom{o}{e}}(t)} \frac{dt}{t}.$$

To understand the L^2 boundedness question, one then needs to understand whether the functions m^e_Γ and m^o_Γ are bounded. The result for H^e_Γ is (assuming $\Gamma(t)$ convex and increasing for $t>0$) the following:

$$\|H^e_\Gamma f\|_{L^2} \le C\|f\|_{L^2}$$

if and only if

(10) $$\Gamma'(At) \ge 2\Gamma'(t)$$

for some $A>1$ and all $t>0$.

The result for H^o_Γ is expressed in terms of a functional

$$h(t) = t\Gamma'(t) - \Gamma(t).$$

Geometrically, $-h(t)$ represents the y intercept of the line tangent to γ at the point $(t,\Gamma(t))$.

The result for H^o_Γ (assuming Γ increasing and convex) is the following

$$\|H^o_\Gamma f\|_{L^2} \le C\|f\|_{L^2}$$

if and only if

(11) $$h(At) \geq 2h(t).$$

See [NVWW1].

It is easy to see that whenever (10) holds, (11) also holds. However, if $\Gamma(t) = \frac{t}{\log\frac{1}{t}}$ for example, (11) holds, but (10) does not. For generalizations to higher dimensions see [NVWW2].

As far as other L^p's go, it has been shown that (if $\Gamma(t)$ is convex and increasing) (10) implies

$$\|H^o_\Gamma f\|_{L^p} \leq C_p\|f\|_{L^p}, \qquad 1<p<\infty,$$

and

$$\|H^e_\Gamma f\|_{L^p} \leq C_p\|f\|_{L^p}, \qquad 1<p<\infty.$$

See [CCCDRVWW]. Thus the L^p boundedness of H^e_Γ is completely understood. It is not known if (11) implies

$$\|H^o_\Gamma f\|_{L^p} \leq C_p\|f\|_{L^p}$$

for $p\neq 2$.

The situation for the maximal function is less clear. It is known that (10) implies-

$$\|\mathcal{M}_\Gamma f\|_{L^p} \leq C_p\|f\|_{L^p} \qquad 1<p\leq\infty$$

[CCCDRVWW].

Moreover, (11) implies

(12) $$\|\mathcal{M}_\Gamma f\|_{L^2} \leq C\|f\|_{L^2}$$

[NVWW3].

An interesting feature of the proof of (12) is the use of the maximal function associated to certain parallelograms. The general technique is the use of a g-function argument in the spirit of [ST]. See also [SW]. But the comparison averages are averages over certain families of parallelograms associated to $\gamma(t)$, as shown in the picture below.

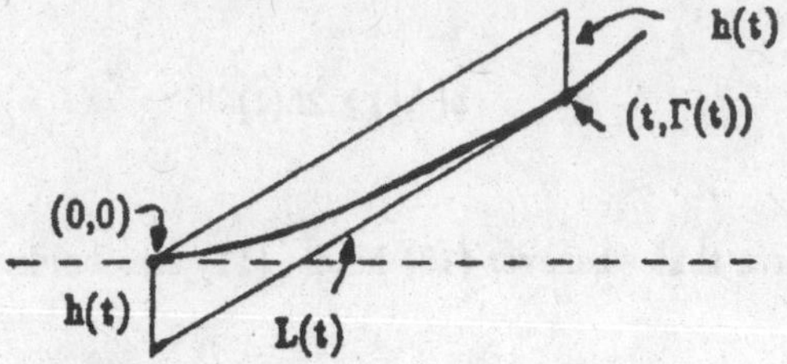

The line $L(t)$ is tangent to γ at $(t,\Gamma(t))$.

It turns out that (11) is not necessary for (12). See [CW]. Finally, Stromberg [STR] has constructed an interesting family of examples to show that (12) does not hold for all convex Γ.

Section 2. A problem in the convergence of Fourier Series.

We turn now to a problem in the convergence of Fourier Series. Suppose ϕ is a smooth homeomorphism from the circle onto the circle. The problem is to determine what conditions on ϕ guarantee that $f(\phi(x))$ has a uniformly convergent Fourier Series whenever f has an absolutely convergent Fourier Series. Strangely, this problem is very similar to the problem of showing that the Hilbert Transform along a curve is bounded on L^2. To see this note that if

$$f(x) = \Sigma\, a_n e^{inx},$$
$$f(\phi(x)) = \Sigma\, a_n e^{in\phi(x)}.$$

Thus if $S_k(g)$ denotes the kth partial sum of the Fourier Series of g,

$$S_k(f(\phi))(x) = \Sigma\, a_n S_k(e^{in\phi(x)}).$$

A little functional analysis then shows $f(\phi(x))$ has a uniformly convergent Fourier Series for every f with absolutely convergent Fourier Series if and only if $S_k(e^{in\phi(x)})$ is uniformly bounded in x, k, and n. Thus the problem reduces to studying

$$\int_{-\pi}^{\pi} \frac{\sin\ k(x-t)}{(x-t)}\, e^{in\phi(t)}\, dt. \tag{13}$$

We now observe that the integral in (13) is very similar to the integral m_ϕ for the

Hilbert Transform along the curve $(t,\phi(t))$. A theorem of Alpar and Kaufman asserts that if

$$(14) \qquad \sum_{j=2}^{m} |\phi^{(j)}(t)| > \delta > 0 \quad \text{(uniformly in t)}$$

for some fixed m, then the integral in (13) is uniformly bounded. In other words (14) implies $f(\phi(x))$ has a uniformly convergent Fourier Series whenever f has an absolutely convergent Fourier Series. Moreoever, ϕ smooth is not a sufficient condition for the boundedness of (13). See [A], [H], [KA] and [K].

The situation in which ϕ has one flat point (say the origin), i.e. $\phi^{(k)}(0)=0$, $k=1,2,3,\ldots$, was studied by Santos [SA]. Santos assumed that for some positive m and each $\delta>0$ there is an $\epsilon(\delta)$ such that

$$(15) \qquad \sum_{j=2}^{m} |\phi^{j}(t)| > \epsilon(\delta)$$

for $|t|>\delta$. He further assumed that

(16–a) $\qquad \phi(0) = \phi'(0) = 0$

(16–b) $\qquad \phi(t)$ is odd near $t=0$

(16–c) $\qquad \phi(t)$ is increasing and convex for $t>0$

(16–d) $\qquad \phi'(At)>2\phi'(t)$ for some $A>1$ and t small and positive.

Under assumptions (15) and (16), Santos found a necessary and sufficient condition that $f(\phi)$ has a uniformly convergent Fourier Series whenever f has an absolutely convergent Fourier Series.

An interesting point is that an extra condition is needed to guarantee the boundedness of the integral in (13). Let us try to see why this is so. In the portion of the integral in (13) for which $|k|\ |x-t|>1$, we might as well write

$$\sin k(x-t) = \frac{1}{2i}\left[e^{ik(x-t)} - e^{-ik(x-t)}\right],$$

and thus consider integrals of the form

$$\int e^{\pm ikt} e^{in\phi(t)} \frac{dt}{(x-t)}.$$

Let us suppose k is positive, n is positive and $\phi'(x) = \frac{k}{n}$, and consider the integral

$$\int e^{-ikt} e^{in\phi(t)} \frac{dt}{x-t}. \tag{17}$$

Then there will be a point $t_1 > x$ so that for $t > t_1$, $n\phi'(t) - k$ will be large enough so that the oscillation of $e^{in\phi(t)-ikt}$ implies the boundedness of $\int_{t>t_1}$. Similarly, there will be a $t_2 < x$ so that for $t < t_2$ $k - n\phi'(t)$ will be large enough so that the oscillation of $e^{in\phi(t)-ikt}$ implies the boundedness of $\int_{t<t_2}$. Now in order for (17) to be bounded, the ratio

$$\frac{t_1 - x}{x - t_2}$$

must be bounded above and below. However, if $\phi'(t)$ is increasing very slowly for $t<x$ and then very rapidly for $t>x$, the ratio will not remain bounded. The condition Santos imposes is a weak form of the assumption that $\log \phi'(t)$ is concave. (Note that in examples like $\phi'(t)=t^{\alpha}$, $\phi'(t)=e^{-1/t^{\beta}}$, $\log \phi'$ is concave.) Precisely, Santos proves the following theorem:

Theorem: If (15) and (16) hold $f\circ\phi$ has a uniformly convergent Fourier Series whenever f has an absolutely convergent Fourier Series if and only if the following condition holds:

(18) There are constants s, M and λ with $0<s<1$, $M>0$. amd $\lambda>1$, such that for $0<a<b$ with a,b small

$$\frac{\phi'(b)-\phi'((1-s)a+sb)}{\phi'((1-s)a+sb)-\phi'(a)} \le M$$

whenever

$$\phi''((1-s)a+sb) \le \lambda\phi'(a).$$

As we implied above (18) holds if $\log \phi'(t)$ is concave for $t>0$.

Section 3. Decay estimates for the Fourier Transform of measures supported on convex hypersurfaces.

If S is an n−1 dimensional hypersurface in R^n, we shall say S is of finite type if any tangent line to S has finite order of contact with S. Let S be a smooth convex n−1 dimensional surface of finite type with surface area measure $d\sigma$. Let $\psi(x)$ be a smooth compactly supported function on S, such that the normal to S at no 2 points of ψ are parallel. Then Bruna, Nagel and Wainger [BNW] have obtained decay estimates for

$(\psi d\sigma)^\wedge$ in terms of the volumes of certain balls on S. For x in S and $t>0$, $B(x,\epsilon)$ consists of those points y in S at a distance of at most ϵ from the hyperplane tangent to S at x. Then the theorem of [BNW] asserts the following:

Theorem: Let S be a convex hypersurface of finite type. Let ξ be a unit vector in R^n. Let x be a point in the support of ψ (x is necessarily unique by the assumption on ψ) such that the normal to S at x is parallel to ξ. Then

$$|(\psi d\sigma)^\wedge(\lambda\xi)| \leq C d\sigma(B(x,\tfrac{1}{\lambda})). \tag{19}$$

The constant C depends only on the support of ψ and the L^∞ norm of a bounded number of derivatives of ψ. See [BNW]. Related results are in [CM]. (If ξ is restricted to a compact set K uniformly bounded away from the support of ψ,

$$|(\psi d\sigma)^\wedge(\xi)| \leq C_m(k)\lambda^{-m}$$

for every m.)

It would seem natural to ask whether (19) might hold if the finite type hypothesis on S is dropped. Bruna, Nagel and Wainger showed that (19) holds for any convex curve in R^2. The question of what happens for surfaces in R^3 has been taken up by Bak, McMichael, Vance, and Wainger [BMVW]. They considered first the case that the surface is the graph of a radial function. They assumed that S is the graph of a radial function. That is

$$S=\{(x,y,z) \mid z=\phi(\sqrt{x^2+y^2})\}$$

and that $\psi=\psi(\sqrt{x^2+y^2})$. They assume ϕ is smooth, ϕ is increasing and convex, and that $\phi(0)=\phi'(0)=0$. They then proved the following theorem:

(20) Theorem: If $\phi(x)$ is as above and

(21) $$\frac{\phi'(r)}{r} \text{ is increasing}$$

then (19) holds.

They also gave examples to show that if the hypothesis (21) is weakened to

$$\frac{\phi'(r)}{r^{1-\epsilon}} \text{ is increasing}$$

then (19) may fail.

The next step might be to assume that S is the graph of a function

$$z = \theta(x,y)$$

such that the level sets of θ coincide with the level sets of a convex polynomial, $p(x,y)$. In this case $\psi(x,y)$ would be a function of $p(x,y)$. Bak, McMichael, Vance and Wainger have made the following step in this direction:

(22) Theorem: Let $p(x,y)$ be a non–negative,even, homogeneous polynomial of degree m such that the set

$$\{(x,y) \mid p(x,y) \leq 1\}$$

is convex. Assume

$$\theta(x,y) = h((P(x,y))^{1/m})$$

with $h(0) = h'(0) = 0$. Let S be the graph of θ, and let ψ be a function of $P^{1/m}$. Then if $\frac{h'(r)}{r}$ is increasing (19) holds.

The proof of (21) depends on expressing the Fourier Transform of $(\psi d\sigma)^\wedge$ in terms of Bessel functions. The proof of (22) rests on a strong form of the result of Bruna, Nagel, Wainger. In [BNW] it is shown that for surfaces of finite type

(23) $$(\psi d\sigma)^\wedge = e^{i\lambda\varphi(\xi)}F(\lambda)$$

where $e^{i\lambda\varphi(\xi)}$ is an explicit oscillatory term, and

(24) $$\left|\frac{d^k F(\lambda)}{d\lambda^k}\right| \leq C_k \frac{d\sigma(B(x,1/\lambda))}{\lambda^k}$$

for each k.

Now to prove (22) one integrates over the level sets of P. Then using the homogeneity of P, one is reduced to an integration in the radial direction and an integration over the surface P=1. One then uses (23) and (24) to replace the standard estimates for Bessel functions. We refer the reader to [BMVW] for the complicated details.

Section 4. Maximal functions on S.

Let S be a compact convex hypersurface in R^n, and let $B(x,\epsilon)$ be the balls defined in section 3. For f in $L^1(S)$ (with measure $d\sigma$= surface area measure), we define

$$\mathcal{M}f(x) = \sup_{\epsilon>0} \frac{1}{d\sigma(B(x,\epsilon))} \int_{B(x,\epsilon)} |f(y)|\, d\sigma(y) .$$

We can then ask if the transformation $f \to \mathcal{M}f$ is of weak type 1–1 or bounded in L^p, $1<p\leq\infty$.

If S is of finite type it follows from properties of the $B(x,\epsilon)$ proven in [BNW] that standard convering lemmas of the Vitali type hold, and one easily sees that the transformation

$$f \to \mathcal{M}f$$

is of weak type one–to–one and bounded on L^p, $1<p\leq\infty$.

Again the question arises as to what happens if the finite type hypothesis is removed. This question was considered by McMichael [M]. He assumes that S is of finite type except at one point, say the origin (which we assume to be a point on S). He assumes that near the origin S is the graph of a radial function. That is

$$S = \{(x,y,z)\} \mid z= \phi(\sqrt{x^2+y^2})\}$$

where $\phi(0) = \phi'(0) = 0$. If for example

$$\phi(x) = e^{-1/x^2} ,$$

the balls $B(x,\epsilon)$ can be long and thin (long in the direction of the origin) and standard properties of balls fail. For example, if

$$d\sigma(B(x,\epsilon)) = d\sigma(B(y,\delta))$$

and

$$B(x,\epsilon) \cap B(y,\delta) \neq \emptyset ,$$

there may not be an η so that

$$B(y,\delta) \subset B(x,\eta)$$

and

$$d\sigma(B(x,\eta) \leq C\, d\sigma(B(x,\delta)).$$

So one cannot employ the standard argument as appears in [STE2]. Nevertheless, McMichael proves the following variant of the Vitali convering lemma:

Lemma: Let $\phi'(x)$ be convex, and assume $\log \phi'(x)$ is concave. Then for any $B(x,\epsilon)$

$$d\sigma\Big(\bigcup_{\substack{B(y,\delta)\cap B(x,\epsilon)\neq\emptyset \\ d\sigma(B(y,\delta))\le d\sigma(B(x,\epsilon))}} B(y,\delta) \Big) \le C\, d\sigma(B(x,\epsilon)) .$$

This lemma allows McMichael to obtain the following theorem:

(25) **Theorem:** With S as above, the mapping

$$f \to \mathcal{M} f$$

is of weak type 1–1.

McMichael also shows that (25) does not follow from the assumption that ϕ is convex.

Section 5. Restriction theorems for the Fourier Transform

The restriction problem is to determine p and q so that

$$(26) \qquad \Big(\int_S |\hat{f}(x)|^q d\sigma\Big)^{1/q} \le C(p,q)\|f\|_{L^p(R^n)}$$

That inequalities of this type hold was first pointed out by Stein. Further progress was made by Fefferman and Stein[F], Sjölin [SJ], Christ [C1], [C2], Drury [D], Drury and Marshall [DM], Tomas [T], Strichartz [STRI], Prestini [P], and Ruiz [R].

We cannot expect inequalities such as (26) to hold if S is flat to infinite order at some point. Nevertheless, one might ask whether there is an Orlicz space version in this case.

This problem was examined by Bak [B]. Bak considered the case of a curve $\gamma(t) = (t,\Gamma(t))$ in the plane with $\Gamma(0)=\Gamma'(0)=0$, with $\Gamma(t)$ increasing and strictly convex on $[0,a)$. To state Bak's result we need the following definition:

Definition: Γ_1 is flatter than Γ_2 if $\dfrac{\Gamma_1''(t)}{\Gamma_2''(t)}$ is non–decreasing.

Then Bak proved the following theorem:

Theorem: Let $\Gamma(t)$ be as above and assume $\Gamma(t)$ is flatter than t^1 on some interval, (c,δ_1), for l=2,3,...

Then for $0<d<1$, $1\le q<\frac{1}{d}$,

(27) $$\left(\int_0^a |\hat{f}(t)\Gamma(t))|^q\, dt\right)^{1/q} \le C_1+C_2\int_{R^2} |f|[\Gamma_e^{-1}(|f|)]^d\, dx.$$

C_1 and C_2 are independent of f and $\Gamma_e(t)$ agrees with $\Gamma(t)$ near t=0. (Γ may not be defined on the range of f.) Moreover the conditions on d and q are sharp (except for the endpoints).

The proof generally follows the lines of [F]. However, a clever idea is needed to measure the precise size of the density of $d\mu*d\mu$ where $d\mu$ is a measure suppported on γ.

Further results and extensions to R^n are in [B].

Section 6. A final remark.

We now add one final remark. Namely, we point out that there are difficulties that may occur in the case of flat curves or surfaces. The first of these difficulties is that (say a curve) might contain a line segment so that the Fourier Transform of a nice measure, $d\sigma$, supported on the entire curve, will not decay at ∞. Another way of putting the difficulty is to say $d\sigma*d\sigma$ will not be absolutely continuous. To the extent that this difficulty has been overcome, one has used Paley Littlewood Theory to "cut out" bad directions in the manner of Nagel, Stein and Wainger "Differentiation in Lacunary Directions". See [NSW].

The second major difficulty is that there seems to be no natural set of dilations. This means that one cannot employ scaling to compare what happens on different pieces of a curve as in [SW] or use dilations to define the balls for maximal functions and Calderon Zygmund Theories. Furthermore, one cannot use dilations to reduce matters to estimates over a compact set of curves as in [BNW].

To illustrate this point, let us consider an integral of the form

$$F(\lambda) = \int \psi(t)e^{i\lambda P(t)}\, dt$$

where P(t) is a polynomial of degree n with positive coefficients and $P(0)=P'(0)=0$, $\psi\in C_0^\infty$. So

$$P(t) = a_2t^2+ \dots a_nt^n \quad, a_j\ge 0, \text{ not all } a_j=0.$$

We want to show

$$|F(\lambda)| \le C\, P^{-1}\left(\tfrac{1}{\lambda}\right). \tag{28}$$

Let us define p by the equation

$$\lambda P(p) = 1.$$

Then $F(\lambda) = p \int \psi(tp)\, e^{i\lambda P(pt)}\, dt$,

and our work is changed to showing

$$G = \int \psi(tp)\, e^{iQ(t)}\, dt \tag{29}$$

is uniformly bounded for all polynomials $Q(t) = \sum_{j=2}^{A} b_j t^j$ with $b_j \ge 0$ and $\Sigma\, b_j^{\,2} = 1$.

A second method of dealing with $F(\lambda)$ is the following:

$$F(\lambda) = \int_0^{P^{-1}(\frac{1}{\lambda})} + \int_{P^{-1}(\frac{1}{\lambda})}^{1}$$

The first integral is clearly at most $P^{-1}(\frac{1}{\lambda})$, while integration by parts shows the second is at most

$$\frac{C}{\lambda} \int_{P^{-1}(\frac{1}{\lambda})} \frac{P''(t)}{[P'(t)]^2}\, dt + \frac{C}{\lambda P'(P^{-1}(\frac{1}{\lambda}))} + \frac{C}{\lambda} \le \frac{C}{\lambda} \frac{1}{P'(P^{-1}(\lambda)}.$$

It is not difficult to see that this implies (28).

The second method works for more general functions. $P(t)$ can be any convex function with $P(0)=P'(0)=0$. The first method restricts the type of convex function $P(t)$ to one that is approximately a polynomial of degree n, but can be applied to more general integrals. For example, the first method shows that if $P(t)$ is a convex polynomial with $P(0)=P'(0)=0$,

$$\left| \int \psi(t) t e^{i\lambda P(t)} \right| \le C\, [P^{-1}(\tfrac{1}{\lambda})]^2 . \tag{30}$$

The second method does not show (30), and in fact (30) is false for a general convex function, P, satisfying $P(0)=P'(0)=0$. The fact that (30) holds allows one to reduce certain two–dimensional problems to one–dimensional integrals in non–flat cases. In particular, the estimates of [BNW] can be reduced to estimates for one–dimensional integrals, while those occurring in [BMVW] can not.

BIBLIOGRAPHY

|A| L. Alpar, "Sur certaines transformées des series de puissance absolument convergentes sur la frontière de leur cercle de convergence," Magyar Tud. Akad. Mat. Kutato Int. Kozl. 7 (1962), pp 287-316.

|A1| L. Alpar, "Sur une classe particulière de Series de Fourier ayant de sommes partielles bornées", Studia Sci. Math. Hungar 1 (1966), pp 189-204.

|B| J.-G. Bak, to appear.

|BMVW| J.-G. Bak, D. McMichael, J. Vance, and S. Wainger, "Fourier Transforms of surface area measure on convex surfaces in R^3", Mathematical Sciences Research Institute Publication, 1988, Berkeley, CA.

|BNW| J. Bruna, A. Nagel, and S. Wainger, "Convex hypersurfaces and Fourier Transforms", Ann. of Math, 127 (1988), pp 333-365.

|CCCDRVWW| H. Carlsson, A. Córdoba, M. Christ, J. Duoandikoetxea, J.L. Rubio de Francia, J. Vance, S. Wainger and D. Weinberg, "L^p estimates for Maximal functions and Hilbert Transforms along flat convex curves in R^2, Bull. of A.M.S. 14 (1986), pp 263-267.

|CW| H. Carlsson and S. Wainger, "Maximal functions related to convex polygonal lines", Ind. Univ. Math. J. 34 (1985), pp 815-823.

|C| M. Christ, "Weak type (1,1) bounds for rough operators", to appear in the Ann. of Math.

|C1| M. Christ, "On the restriction of the Fourier Transform to curves: End-point results and the degenerate case", Trans. A.M.S. 287 (1985), pp 223-238.

|C2| M. Christ, "Restriction of the Fourier Transform to submanifolds of low co-dimension", Ph.D. Thesis, University of Chicago.

|C3| M. Christ, "Hilbert transforms along curves II: A flat case", Duke Math. J. 52(1985) pp 887-894.

|CS| M. Christ and E. Stein, "A remark on singular Calderón-Zygmund Theory", Proc. of A.M.S., 99 (1987), pp 71-75.

|CO| A. Córdoba, A. Nagel, J. Vance, S. Wainger and D. Weinberg, "L^p bounds for Hilbert Transforms along convex curves", Inventiones Math. 83 (1986) pp 59-71.

|CM| M. Cowling and G. Mauceri, "Oscillatory integrals and Fourier transforms of surface carried measures", Trans. Amer. Math. Soc. 304 (1987), pp 53-68.

|D| S. Drury, "Restrictions of Fourier transforms to curves", Ann. Inst. Fourier Grenoble 35 (1985), pp 117-123.

|DM| S. Drury and B. Marshall, "Fourier restriction theorems for curves with affine and Euclidean arclengths", Math. Proc. Cambridge Philos. Soc. 97 (1985), pp 111-125.

|F| C. Fefferman, "Inequalities for strongly singular convolution operators", Acta. Math. 75 (1970), pp 9-36.

|H| G. Halasz, "On a theorem of L. Alpar concerning Fourier series of certain functions", Stud. Sci. Math. Hungarica 2 (1967), pp 67-72.

|KA| R. Kaufman, "Uniform convergence of Fourier series", Stud. Sci. Math. Hungarica, 10 (1975), pp 81-83.

|K| J.P. Kahane, "Quatre Leçons sur les Homeomorphisms du Cercle et les Series de Fourier", Topic in Harmonic Analysis, Instituto Nazionale di Alta Matematica, Rome (1983), pp 955-990.

|M| D. McMichael, to appear.

|NSW| A. Nagel, E.M. Stein, and S. Wainger, "Differentiation in Lacunary directions", Proc. Nat. Acad. Sci. USA 75 (1978), pp 1060-1061.

|NVWW1| A. Nagel, J. Vance, S. Wainger, and D. Weinberg, "Hilbert Transforms for convex curves", Duke Math. J. 50 (1983), pp 735-744.

|NVWW2| A. Nagel, J. Vance, S. Wainger, and D. Weinberg, "The Hilbert Transform for convex curves in R^n", Amer. J. Math. 108 (1986), pp 485-504.

|NVWW3| A. Nagel, J. Vance, S. Wainger and D. Weinberg, "Maximal functions for convex curves", Duke Math. J. 52 (1985), pp 715-722.

|P| E. Prestini, "Restriction theorems for the Fourier Transform to some manifolds in R^n", Proc. Sympos. Pure Math. 35 Part 1, A.M.S. Providence (1979), pp 101-109.

|R| A. Ruiz, "On the restriction of Fourier Transforms to curves", Conf. Harmonic Analysis in Honor of A. Zygmund Vol. I, Wadsworth, Belmont, CA (1983), pp 186-212.

|SA| A. Santos, Ph.D. Thesis at University of Wisconsin, June 1987.

|SJ| P. Sjölin, "Fourier multipliers and estimates of Fourier transforms of measures carried by smooth curves in R^2", Studia Math. |51| (1974), pp 169-182.

|STE| E. Stein, "Maximal functions: Homogeneous curves", Proc. Nat. Acad. Sci. U.S.A. 73, pp 2176-2177.

|STE2| E. Stein, <u>Singular Integrals and Differentiability Properties of Functions</u>, Princeton Univ. Press, Princeton (1970).

|SW| E. Stein and S. Wainger, "Problems in harmonic analysis related to curvature", Bull. of A.M.S., 84 (1978), pp 1239-1295.

|STRI| R. Strichartz, "Restrictions of Fourier Transforms to quadratic surfaces and decay of solutions of wave equations", Duke J. Math. 44 (1977), pp 705-714.

|STR| J.O. Stromberg, personal communication.

|T| P. Tomas, "A restriction theorem for the Fourier Transform", Bull. Amer. Math. Soc. 81 (1975), pp 477-478.

SOME TOPICS IN SYMMETRIZATION

Albert Baernstein II[1]
Mathematics Department, Washington University
St. Louis, Missouri, 63130, U.S.A.

Suppose that u is a real valued measurable function on $\mathbb{R}^n$. The function u_0 defined by

$$u_0(x) = \inf\left\{t: |u > t| \leq \omega_n |x|^n\right\},$$

where $|E|$ denotes the Lebesgue measure of E and ω_n the measure of the unit ball, is known variously as the symmetric decreasing rearrangement, the Schwarz symmetrization, or the point symmetrization of u. It exists provided $|u > t| = |\{x \in \mathbb{R}^n: u(x) > t\}|$ is finite for every $t > \operatorname{ess\,inf}_{\mathbb{R}^n}$, is a radial function which decreses when $|x|$ increases, and satisfies $|u > t| = |u_0 > t|$ for every t.

Next, we consider for $x \in \mathbb{R}^n$ the spherical coordinates r, θ defined by $r = |x|$, $x_1 = r \cos \theta$, $0 \leq \theta \leq \pi$. We will not need to consider the other angular coordinates. Let S^{n-1} denote the unit sphere $\{x \in \mathbb{R}^n: r = 1\}$, $C(\theta_0)$ the spherical cap on S^{n-1} given by $C(\theta_0) = \{x \in S^{n-1}: 0 \leq \theta < \theta_0\}$ and $d\sigma$ the surface area measure on S^{n-1}. If u is a measurable function on S^{n-1} its spherical symmetrization $u_0: S^{n-1} \to \mathbb{R}$ can be defined as in the point symmetrization case. u_0 is a function of θ, decreases with θ, and satisfies $\sigma(u > t) = \sigma(u_0 > t)$ for every t. If u is defined in the spherical shell $A(R_1, R_2) = \{x \in \mathbb{R}^n: R_1 < |x| < R_2\}$ its spherical symmetrization $u_0 = u_0(r, \theta)$ is defined by symmetrizing on each sphere $|x| = r$, $R_1 < r < R_2$.

There are many other types, for example k-dimensional cap symmetrizations, which include spherical and "cylindrical" symmetrizations, and k-dimensional Steiner symmetrizations,

1 This research was supported in part by a grant from the National Science Foundation.

which involve symmetrizing in a plane of dimension $n - k$ to obtain a u_0 which depends on $n - k + 1$ variables. An account of these can be found in [Sa]. As part of an attempt to make this article readable, we will only discuss the point and spherical cases, but all the results we mention have cap and Steiner analogues.

The passage from u to u_0 has been the key step in the solution of extremal problems from many different parts of analysis. One notable result, with many applications, is that symmetrization decreases the Dirichlet integral. If $u:\mathbb{R}^n \to \mathbb{R}$ is Lipschitz and satisfies $\lim_{x\to\infty} u(x) = \inf_{\mathbb{R}^n} u$, then for point symmetrization we have

$$\text{(DI)} \qquad \int_{\mathbb{R}^n} |\nabla u|^p \, dx \geq \int_{\mathbb{R}^n} |\nabla u_0|^p \, dx, \qquad 1 \leq p < \infty.$$

For Lipschitz functions u on $A(R_1,R_2)$ and spherical symmetrization, such an inequality holds on each sphere $|x| = r$, with dx replaced by $d\sigma$.

The case $p = 2$ of point symmetrization seems to have first appeared in independent papers by Faber and Krahn in the 1920's, in which they proved Lord Rayleigh's conjecture that among all fixed membranes in the plane with given area the disk has the smallest fundemental frequency. Their argument works just as well in $\mathbb{R}^n$. For a domain $\Omega \subset \mathbb{R}^n$ the first eigenvalue for the problem $\Delta u + \lambda u = 0$, $u = 0$ on $\partial\Omega$ is given by the Rayleigh quotient

$$\inf \frac{\int_\Omega |\nabla u|^2}{\int_\Omega u^2}$$

where the $\inf$ may be taken over all Lipschitz functions which vanish near $\partial\Omega$. Since $\left|\nabla|u|\right| \leq |\nabla u|$, we may restrict attention to nonnegative u. Then, setting $u = 0$ in $\mathbb{R}^n\setminus\Omega$, we obtain a function satisfying the hypotheses of (DI). For the point symmetrization u_0 we have $\int u^2 = \int u_0^2$, and the inequality $\lambda_1(\Omega) \geq \lambda_1(\Omega_0)$ easily follows.

The DI inequality was proved for some symmetrizations and applied to various other "isoperimetric" inequalities of physics in the 1951 book of Pólya and Szegö [PS]. A good survey of this subject, as it existed in 1967, can be found in [Pa]. Hayman [Ha] used the case $p = 2$, $n = 2$ for Steiner and spherical symmetrization to prove theorems about growth of analytic functions which generalize Schwarz' lemma. For $p = n = 3$ and spherical symmetrization (DI) was essentially proved by Gehring [Ge] in 1961 as a key ingredient in his theory of quasiconformal maps in space. The passage to $\mathbb{R}^n$, $n \geq 4$, again with $p = n$, was carried out by Mostow [Mo]. The first complete proofs of (DI) were published by Sperner [Sp 1], [Sp 2]. They rely on general forms of the isoperimetric inequality in $\mathbb{R}^n$ and S^{n-1}, and make extensive use of geometric measure theory. Uniqueness questions associated with (DI) have been a topic of much recent research. Some of the best results appear in [BZ].

Another general inequality concerns the convolution of three functions. Suppose that f,g,h are nonnegative functions on $\mathbb{R}^n$. Then, for point symmetrization, we have

$$\text{(TC)} \quad \int_{\mathbb{R}^n}\int_{\mathbb{R}^n} f(x)g(y)h(x-y)\,dxdy \leq \int_{\mathbb{R}^n}\int_{\mathbb{R}^n} f_0(x)g_0(y)h_0(x-y)\,dxdy.$$

For $n = 1$ this inequality is due to F. Riesz, 1930, and for general n to Sobolev, 1938. An extension involving more than three functions appears in [BLL], and interesting related work, with an application to eigenfunctions, is in [BL].

A spherical version of (TC) was proved in [BT]. Suppose that f,g are functions in $L^1(S^{n-1})$ and that $K(t) \in L^\infty[0,2]$ is a nonincreasing function of t. Then, with the 0 denoting spherical symmetrization,

$$\text{(STC)} \quad \int f(x)g(y)K(|x-y|)\,d\sigma(x)d\sigma(y) \leq \int f_0(x)g_0(y)K(|x-y|)\,d\sigma(x)d\sigma(y)$$

where the integration is taken over $S^{n-1} \times S^{n-1}$ and $|x - y|$ is the distance in $\mathbb{R}^n$. Recently, I have figured out how to drop the monotonicity assumption on the third function when $n = 1$, and in a forthcoming paper [B5] will prove that

$$\int_{-\pi}^{\pi}\int_{-\pi}^{\pi} f(e^{i\theta})g(e^{i\varphi})h(e^{i(\theta-\varphi)})\,d\varphi\,d\theta$$

increases under circular symmetrization when, for example $f, g \in L^1(S^1)$, $h \in L^\infty(S^1)$. Presumably, a higher dimensional analogue exists for convolutions on $SO(n)$, but my proof only works on the circle.

Recall that $A(R_1, R_1) = \{x \in \mathbb{R}^n : R_1 < |x| < R_2\}$. Let $A^*(R_1, R_2) = (R_1, R_2) \times (0, \pi)$, which we view as a subset of the upper half plane in $\mathbb{R}^2$. Define an operator J, which takes functions on $A = A(R_1, R_2)$ to functions on A^*, by

$$Ju(r,\theta) = \int_{C(\theta)} u(rx)\,d\sigma(x),$$

and for $u \in L^1(A)$ define u^* on A^* by

$$u^*(r,\theta) = Ju_0(r,\theta) = \sup \int_E u(rx)\,d\sigma(x),$$

where the sup is taken over all $E \subset S^{n-1}$ with $\sigma(E) = \sigma(C(\theta))$. A proof of the second equation may be found in [B1].

In spherical coordinates the Laplace operator is $\Delta = r^{1-n} \frac{\partial}{\partial r}\left[r^{n-1} \frac{\partial}{\partial r}\right] + r^{-2}\left[\sin\theta\right]^{2-n} \frac{\partial}{\partial\theta}\left[\sin^{n-2}\theta \frac{\partial}{\partial\theta}\right]$. Let L denote the operator

$$L = r^{1-n} \frac{\partial}{\partial r}\left[r^{n-1} \frac{\partial}{\partial r}\right] + r^{-2}\left[\sin\theta\right]^{n-2} \frac{\partial}{\partial\theta}\left[\sin^{2-n}\theta \frac{\partial}{\partial\theta}\right].$$

Writing $d\sigma(x)$ in spherical coordinates, it's easy to check that $LJ = J\Delta$.

In [BT], Taylor and I proved

_If_ u _is subharmonic in_ A, _then_ $Lu^* \geq 0$, _in the sense of distributions, on_ A^*.

I had earlier proved the case $n = 2$. Gariepy and Lewis [GL] also proved a version of the n dimensional case. There have been applications to, among other things, meromorphic functions ("spread relation" and improved defect relation),

Riemann mapping functions (the Koebe function has maximal integral means), and potential theory (symmetrization "increases" Green functions and harmonic measures). A list of papers upto 1978 appears in the survey [B2].

The fact that u subharmonic implies u^* is L-subharmonic is actually a special case of a general heuristic principle. For fixed $(r,\theta) \in A^*$ let $E \subset S^{n-1}$ be a set with $\sigma(E) = \sigma(C(\theta))$ for which the sup in the definition of u^* is attained. Such a set is the union of some set $\{x: u > t\}$ with an appropriately sized subset of $\{x: u = t\}$. Assume that u is the difference of two subharmonic functions, so that Δu is a locally finite signed measure on A.

$$(*P) \qquad \int_E \Delta u(rx)\, d\sigma(x) \le \int_{C(\theta)} \Delta u_0(rx)\, d\sigma(x) = Lu^*(r,\theta).$$

The equality is a consequence of the relation $J\Delta = LJ$. It is not clear what the inequality means precisely, since u and u_0 can be nonsmooth. A precise result of this type is formulated below as Theorem 1. A formal "derivation" of (*P) follows easily from (STC) if we write

$$\Delta u(x) = \lim_{\delta\to 0} C_n \delta^{-2}\left[\left[\omega_n^{-1}\delta^{-n}\int_{|y|<\delta} u(x+y)\,dy\right] - u(x)\right]$$

and express the mean value in terms of spherical convolutions.

The form of the following theorem is motivated by the applications we wish to make to linear and nonlinear p.d.e. Assume that u is the difference of two subharmonic functions in A, that φ is a Borel measurable function defined on the range of u for which $\varphi \circ u \in L^1_{loc}(A)$, and that μ is a locally finite signed measure on A. Then $\mu = \rho dx + \tau^+ - \tau^-$ where $\rho \in L^1_{loc}(A)$ and τ^+, τ^- are positive singular measures. Define measures μ_0 on A, μ^* on A^*, by $\mu_0 = \rho_0 dx + \tau_0^+$, $\mu^* = \rho^* dx + \tau^*$. Here τ_0^+ is the measure concentrated on the positive x_1 axis which satisfies $\tau_0^+\left[A(R_3,R_4)\right] = \tau^+\left[A(R_3,R_4)\right]$ for every $R_1 < R_3 < R_4 < R_2$, and τ^* is obtained by "integrating" τ_0^+: For $F \in C_c(A^*)$,

$$\int_{A^*} F\, d\tau^* = \int_{R_1}^{R_2} r^{2-n}\, d\tau_0^+(r) \int_0^{\pi} F(r,\theta)d\theta.$$

The dx in the definition of μ^* refers to Lebesgue measure on $\mathbb{R}^2$, and the r^{2-n} appears in the definition of τ^* so that if μ is approximated by absolutely continuous measures $h\,dx$ then μ^* is approximated by h^*dx.

<u>Theorem 1</u>. <u>If</u> $\Delta u + \varphi(u) + \mu = 0$ <u>in</u> A, <u>then</u>

$$Lu^* + J\varphi(u_0) + \mu^* \geq 0 \quad \text{in} \quad A^*.$$

The equation and inequality take place in the sense of distributions. Taking $\varphi = 0$ $\mu = -\Delta u$, we deduce that subharmonicity of u implies L subharmonicity of u^*. The proof of this in [GL] used geometric measure theory, while that of [BT] was based on (STC) but was rather complicated. In principle, the proof of the more general Theorem 1 follows in straightforward fashion from (STC) together with the construction of an appropriate measure preserving transformation T for which $u = u_0 \circ T$. A transformation of this type appears also in [BZ]. Details of the proof of Theorem 1 will appear in [B6].

A version of Theorem 1 still holds when the full shell A is replaced by a domain $\Omega \subset A$, provided u satisfies an appropriate condition on the "interior" boundary $(\partial\Omega) \cap A$. A typical sufficient condition is $u > 0$ in Ω and $u = 0$ on the interior boundary. This makes it possible to prove comparison theorems for p.d.e.'s. A full account will appear in [B6]. Here we give a sample.

Let φ be as in Theorem 1 and μ be a locally finite measure on Ω. Assume also that for each domain $\Omega' = \Omega \cap A(R_3,R_4)$ with $R_1 < R_3 < R_4 < R_2$ we have $u \circ \varphi \in L^1(\Omega')$, and that μ has finite total variation on each Ω'. Let Ω_0 be the domain obtained from Ω by spherical symmetrization. It is a union of spherical caps centered on the positive x_1 axis, and the symmetrized measure μ_0 can be defined on Ω_0. Suppose that u,v are continuous and positive in Ω,Ω_0, repectively, u vanishes on the interior boundary of Ω, v is spherically symmetric and each can be expressed as the difference of two subharmonic functions.

Theorem 2. Suppose also that

(a) $\Delta u + \varphi(u) + \mu \geq 0$ in Ω, $\quad \Delta v + \varphi(v) + \mu_0 \leq 0$ in Ω_0,

(b) $u^* \leq v^*$ on $(\partial\Omega^*) \cap (|x| = R_i)$, $\quad i = 1,2$,

(c) φ is convex and decreasing on $(0,\infty)$.

Then $u^* \leq v^*$ in Ω^*.

Here Ω^* is the subset of $A^* \subset \mathbb{R}^2$ defined by $\Omega^* = \{(r,\theta): 0 < \theta < \theta_r\}$, where $\Omega_0 \cap (|x| = r) = C(\theta_r)$. If $R_1 = 0$ the inequality in (b) comes for free as a consequence of the other hypotheses. The same is true for $R_2 = \infty$ under appropriate Phragmén-Lindelöf type assumptions.

The conclusion $u^* \leq v^*$ can be stated in a more familiar fashion: For every $r \in (R_1,R_2)$ and every convex increasing function Φ,

$$\int_{\Omega_r} \Phi(u(rx))\,d\sigma(x) \leq \int_{C(\theta_r)} \Phi(v(rx))\,d\sigma(x),$$

where $\Omega_r = \{x \in S^{n-1} : rx \in \Omega\}$. Equivalence of two conditions of this type goes back to the book of Hardy, Littlewood, and Pólya. A proof in the present context appears in [B1]. In particular, Theorem 2 asserts that for every $r \in (R_1,R_2)$ and $1 \leq p < \infty$,

$$\int_{\Omega_r} u^p(rx)\,d\sigma(x) \leq \int_{C(\theta_r)} v^p(rx)\,d\sigma(x), \qquad \max_{\Omega_r} u \leq \max_{C(\theta_r)} v.$$

Theorem 2, with slightly stronger regularity hypotheses, was essentially proved by Weitsman [W], who developed an ad hoc version of (*P). The general idea of the proof is to consider $w = u^* - v^*$ in Ω^*, and take $x_0 \in \overline{\Omega^*}$ where w achieves its supremum. By Theorem 1, we have $Lu^* \geq -\mu^* - J\varphi(u_0)$, and equality holds for v^*. Hence $Lw \geq J\varphi(v)$ $J\varphi(u_0)$. Weitsman cleverly shows that if $x_0 \in \Omega^*$ and $w(x_0) > 0$ then $J\varphi(v) - J\varphi(u_0) > 0$ near x_0, which violates the maximum principle for the elliptic operator L. Hence $x_0 \in \partial\Omega^*$, and further analysis shows that $w(x_0) > 0$ is impossible there as well.

If, in Theorem 2, we take $\varphi = 0$ and μ a point mass at $x_0 \in \Omega$, then u is the Green function of Ω with pole at x_0, v the Green function of Ω_0 with pole at $(|x_0|,0,\ldots,0)$, and we recover the results in [Bl], [BT]. Weitsman was led to his theorem by his researches on the Poincaré metric for plane domains. Suppose that $\Omega \subset \mathbb{C}$ has at least two finite boundary points. Then there is a function $\rho:\Omega \to \mathbb{R}^+$ which satisfies

$$\Delta(\log \rho) + \rho^2 = 0, \qquad \lim_{x\to x_0} \rho(x) = \infty.$$

for each finite boundary point x_0.

The maximal such ρ is called the Poincaré metric, and is denoted ρ_Ω. A longstanding conjecture of geometric function theory asserted that

$$\text{(P.M.)} \qquad \inf_{|x|=r} \rho_{\Omega_0}(x) \le \inf_{|x|=r} \rho_\Omega(x), \qquad R_1 < r < R_2,$$

where Ω_0 is the circularly symmetrized domain. For simply connected Ω this inequality is equivalent to the corresponding one for Green functions, and had been proved by Hayman in the 1950's. But for multiply connected Ω the P.M. problem is genuinely nonlinear, and no one could figure out how to deal with it before Weitsman. Set $u = \log \rho_\Omega^{-1}$. Then $\Delta u + e^{-2u} = 0$. From a slightly modified version of Theorem 2, with $\mu = 0$, $\varphi(x) = e^{-2x}$, it follows that not only is (P.M.) true, but in fact all convex integral means of $\log \rho_\Omega^{-1}$ are dominated by those of $\log \rho_{\Omega_0}^{-1}$ on every circle.

We have discussed $*$-functions in some detail for spherical symmetrization in $\mathbb{R}^n$, but the theory works just as well for various other kinds too, such as Steiner and point. Let us return to point symmetrization. Suppose that $u \in L^1_{loc}(\mathbb{R}^n)$ and essentially approaches its inf at ∞, that is

$$\lim_{R\to\infty} \left|\left\{x \in \mathbb{R}^n : |x| > R, u(x) \ge A\right\}\right| = 0$$

for every $\lambda >$ essinf u. The point symmetrization u_0 is then defined on $\mathbb{R}^n$, and we define $u^*:(0,\infty) \to \mathbb{R}$ by

$$u^*(r) = \int_{B(r)} u_0\,dx = \sup \int_E u\,dx$$

where the sup is over all $E \subset \mathbb{R}^n$ with $|E| = |B(r)|$, and $B(r) = \{x: |x| < r\}$.

The analogue of theorem 1 asserts that if u satisfies a p.d.e. $\Delta u + \varphi(u) + \mu = 0$ then u^* satisfies an ordinary differential inequality $Lu^* + \int_{B(r)} \varphi(u)dx + \mu^* \geq 0$, where L is the adjoint of the radial Laplacian, $L = r^{n-1} \frac{d}{dr}\left[r^{1-n}\frac{d}{dr}\right]$. From this follow results like Theorem 2, in which one compares solutions of $\Delta u + \varphi(u) + \mu = 0$ in a domain Ω with solutions of $\Delta v + \varphi(v) + \mu_0 = 0$ in the ball Ω_0 with $|\Omega| = |\Omega_0|$. In the point symmetrization context such theorems go back to Weinberger, 1962, for the case $\varphi = 0$. Later work has been carried out by Bandle, Talenti, and many other authors. Accounts and references may be found in the books [Ban], [Kaw] and [Moss]. It's known, for example, that $u^* \leq v^*$ holds when φ is an increasing function and v is positive in Ω_0. If, in addition, φ is convex, Theorem 1 can be used to prove this for spherical and other symmetrizations. It is not known if monotonicity alone, without the convexity assumption, is enough when the symmetrized function depends on more than one variable. Also, in the point case inequalities between ∇u and ∇u_0 have been proved which go beyond anything done so far in the several variable situation. On the other hand, the comparison theorems for parabolic equations $\Delta u + \varphi(u) + \mu = \frac{\partial u}{\partial t}$ with, say, φ convex decreasing, follow from Theorem 1 in any number of variables. One assumes that $u^*(\cdot,0) \leq v^*(\cdot,0)$, and proves that $u^*(\cdot,t) \leq v^*(\cdot,t)$ for every t.

The proofs in the point case have generally involved analysis on the level sets of u, where differential inequalities are established which play the same role as the inequality $Lu^* + \int_{B(r)} \varphi(u)dx + \mu^* \geq 0$. The level set methods generally require geometric measure theory, and appear to be less easily adaptable to the several variable case than the *-function approach.

In addition to the work discussed above, which applies to point, spherical, Steiner and cap symmetrization, there have been recent advances also on other fronts. C. Borell [Bo] has studied symmetrization in Gauss space, and has proved comparison theorems for parabolic equations involving the operator $\Delta - x\cdot\nabla$ which is attached to the Ornstein-Uhlenbeck process. Berard [Be] and collaborators have defined a cigar shaped symmetrization for Riemannian manifolds, and have used it to prove lower bounds for Laplace-Beltrami eigenvalues and upper bounds for heat kernels. Back in the Euclidean case, Dubinin [Du], see also [B3], [B4], has introduced a new idea, desymmetrization, which he applied to solve a well known conjecture of Gončar about maximizing the harmonic measure $\omega(0,E,\Omega)$, where E is the union of n radial slits $\{re^{i\theta_j}: r_0 \le r \le 1\}$ in the unit disk Δ, and $\Omega = \Delta\backslash E$. The multiply connected and higher dimensional versions of Dubinin's work provide interesting directions for future research.

Many problems about symmetrization remain unsolved. Probably the most famous is the one about the first eigenvalue of the clamped plate. For the eigenvalue problem

$$\Delta\Delta u = \lambda u \quad \text{in} \quad \Omega, \qquad u = \nabla u = 0 \quad \text{on} \quad \partial\Omega,$$

is it true that, as in the membrane case, $\lambda_1(\Omega)$ is minimal among all domains of the same measure when Ω is a ball? This isoperimetric inequality was conjectured also by Rayleigh in the 19'th century. The best known result, $\lambda_1(\Omega_0) \le 1.02\,\lambda_1(\Omega)$, is due to Talenti [T]. The corresponding variational formula is

$$\lambda_1(\Omega) = \inf \frac{\int_\Omega (\Delta u)^2}{\int_\Omega u^2}, \qquad u = \nabla u = 0 \quad \text{on} \quad \partial\Omega.$$

If the first eigenfunction were always positive, then $\lambda_1(\Omega) \ge \lambda_1(\Omega_0)$ would follow from the known technology, but there exist domains for which this function changes sign.

Another classical problem, from complex variables, see [A] or [Hi], is that of finding the exact value of Landau's constant. If f is holomorphic in the unit disk Δ, then Landau proved that $f(\Delta)$ contains a disk of radius $\geq L|f'(0)|$, where L is an absolute constant. It is known that $0.5 \leq L \leq \frac{3}{2\Gamma(\frac{1}{3})} = 0.56\ldots$, and conjectured that the number on the right provides the best value. An equivalent formulation can be phrased in terms of Poincaré metrics. Let Ω_0 be $\mathbb{C}\backslash\mathcal{L}$, where $\mathcal{L}$ is a lattice of equilateral triangles. Suppose that Ω is another domain, which we think of as the complement of a perturbed lattice, and that

$$\sup_{\Omega} \operatorname{dist}(z,\partial\Omega) \leq \sup_{\Omega_0} \operatorname{dist}(z,\partial\Omega_0).$$

Is it true that

$$\inf_{\Omega_0} \rho_{\Omega_0}(z) \leq \inf_{\Omega} \rho_{\Omega}(z)?$$

This is a problem of the same type as that solved by Weitsman. The *-function machine would probably work, if one could surmount the conceptual difficulty of deciding how functions in Ω can be symmetrized to get new ones on Ω_0.

Other symmetrization questions arise in quasiconformal mapping. In dimension two, one would like to know how to compare solutions of Beltrami equtions $f_{\bar{z}} = \mu f_z$ with those of $g_{\bar{z}} = \mu_0 g_z$, where μ_0 is somehow obtained from μ. This provides one means of attacking the "area distortion problem", which is dicussed, for example, in [BM]. More generally, what connections exist between symmetrization techniques and $\bar{\partial}$ problems, or problems in several complex variables?

At the beginning of this article we discussed Dirichlet integral inequalities, and then took up triple convolutions and *-functions. It appears that many of these inequalities in fact have a common source, in an integral inequality of the form

$$\text{(I)} \quad \iint_{|x-y|\leq\delta} \Phi(|u(x)-v(y)|)\,dxdy \geq \iint_{|x-y|\leq\delta} \Phi(|u_0(x)-v_0(y)|)\,dxdy$$

where Φ is convex and increasing. Such inequalities are well known when u and v are functions of one variable, see, e.g. [Gar], and in the several variable case appear in a not yet published manuscript by W. Beckner [Bec]. It should be possible to develope a whole theory of symmetrization ab ovo from (I). I hope to carry out this project sometime in the future.

References

[A] L.V. Ahlfors, Conformal Invariants, McGraw-Hill, New York, 1973.

[B1] A. Baernstein, Integral means, univalent functions, and circular symmetrization, Acta. Math. 133 (1974), 139-169.

[B2] ——————, How the *-functions solves extremal problems, Procedings of the I.C.M., Helsinki, 1978, pp. 639-644, Academia Scientiarum Fennica, Helsinki, 1980.

[B3] ——————, Dubinin's symmetrization theorem, Complex Analysis I, Proceedings, Univ. of Maryland 1985-86, pp. 23-30, C.A. Berenstein, ed., Lecture Notes in Mathematics 1275, Springer-Verlag, Berlin-Heidelberg, 1987.

[B4] ——————, On the harmonic measure of slit domains, to appear in Complex Variables.

[B5] ——————, Convolution and rearrangement on the circle, preprint.

[B6] ——————, untitled work in progress.

[BM] A. Baernstein and J. Manfredi, Topics in quasiconformal mapping, Topics in Modern Harmonic Analysis, pp. 819-862, L. de Michele and F. Ricci, eds., INDAM Francesco Severi, Roma, 1983.

[BT] A. Baernstein and B.A. Taylor, Spherical rearrangements, subharmonic functions, and *-functions in n-space, Duke Math. J. 43 (1976), 245-268.

[Ban] C. Bandle, Isoperimetric Inequalities and Applications, Pitman, Boston, 1980.

[Bec] W. Beckner, preprint.

[Be] P. Bérard, Spectral Geometry: Direct and Inverse Problems, Lecture Notes in Mathematics 1207, Springer-Verlag, Berlin-Heidelberg, 1986.

[Bo] C. Borell, Geometric bounds on the Ornstein-Uhlenbeck velocity process, Z. Wahrscheinlichkeitstheorie verw. Gebiete 70 (1985), 1-14.

[BL] H.J. Brascamp and E. Lieb, On extensions of the Brunn-Minkowski and Prékopa-Leindler theorems, including inequalities for log-concave functions, and with an application to the diffusion equation, J. Funct. Anal. 22 (1976), 366-389.

[BLL] H.J. Brascamp, E. Lieb, and J.M. Luttinger, A general rearrangement inequality for multiple integrals, J. Funct. Anal. 17 (1974), 227-237.

[BZ] J. E. Brothers and W.P. Ziemer, Minimal rearrangements of Sobolev functions, Indiana University preprint, 1987.

[Du] V.N. Dubinin, On the change of harmonic measure under symmetrization (Russian), Mat. Sbornik 124 (1984), 272-279. English translation, Mathematics of the USSR Sbornik 52 (1985), 267-273.

[GL] R. Gariepy and J.L. Lewis, A maximum principle with applications to subharmonic functions, Ark. Mat. 12 (1974), 253-266.

[Gar] A.M. Garsia, Some combinatorial methods in real analysis, Proceedings of the I.C.M., Helsinki, 1978, pp. 615-622, Academia Scientiarum Fennica, Helsinki, 1980.

[Ge] F.W. Gehring, Symmetrization of rings in space, Trans. Amer. Math. Soc. 101 (1961), 499-519.

[Ha] W.K. Hayman, Multivalent Functions, Cambridge University Press, Cambridge, 1967.

[Hi] E. Hille, Analytic Function Theory, Vol. II, Ginn, Boston, 1962.

[Kaw] B. Kawohl, Rearrangements and Convexity of Level Sets in PDE, Lecture Notes in Mathematics 1150, Springer-Verlag, Berlin-Heidelberg, 1985.

[Moss] J. Mossino, Inégalités Isopérimétriques et Applications en Physique, Hermann, Paris, 1984.

[Mos] G. Mostow, Quasi-conformal mappings in n-space and the rigidity of hyperbolic space forms, IHES Publ. Math. 34 (1968), 53-104.

[PS] G. Pólya and G. Szegö, Isoperimetric Inequalities in Mathematical Physics, Princeton Univ. Press, Princeton, 1951.

[Pa] L.E. Payne, Isoperimetric inequalities and their applications, SIAM Rev. 9 (1967), 453-488.

[Sa] J. Sarvas, Symmetrization of condensers in n-space, Ann. Acad. Sci. Fennicae Ser. A I No. 522 (1972), 44 pp.

[Sp1] E. Sperner, Jr., Zur Symmetrisierung von Funktionen auf Sphären, Math Z. 134 (1973), 317-327.

[Sp2] ———, Symmetrisierung für Funktionen mehrerer reeler Variablen, Manuscripta Math. 11 (1974), 159-170.

[T] G. Talenti, On the first eigenvalue of the clamped plate, Ann. Mat. Pura Appl. (4) 129 (1981), 265-280.

[W] A. Weitsman, Symmetrization and the Poincaré metric, Annals of Math. 124 (1986), 159-169.

ON THE SIZE THE MAXIMAL FUNCTION AND THE HILBERT TRANSFORM

JOAQUIM BRUNA

Universitat Autònoma de Barcelona

Bellaterra, 08193 Barcelona, SPAIN

1. Introduction. In this note we present a survey of recent results related to the question of describing the size of the Hilbert transform of L^1 functions or measures. Most of the results quoted here have been obtained in [2] and [6].

The setting will be the following. By $\mathbf{D}$ we will denote the unit disc in the complex plane, $\mathbf{T}$ its boundary and Σ will denote the class of all holomorphic functions f in $\mathbf{D}$ with positive real part, $\operatorname{Im} f(0) = 0$. Each $f \in \Sigma$ has a Herglotz representation

$$f(z) = \frac{1}{2\pi}\int_0^{2\pi} \frac{e^{it}+z}{e^{it}-z}\, d\mu(e^{it}),$$

where μ is a positive measure on $\mathbf{T}$, and has boundary values given a.e. by $f(e^{it}) = \varphi(e^{it}) + i\tilde{\mu}(e^{it})$, where $\varphi\, dt$ is the absolutely continuous part of $d\mu$ and $\tilde{\mu}$ is the conjugate of μ.

We introduce now the class $\mathbf{X}$ as follows:

$$\mathbf{X} = \{\phi : \mathbf{T} \longrightarrow \mathbf{R};\ \exists f \in \Sigma \text{ s.t. } |\phi| \le |f| \text{ a.e. on } \mathbf{T}\}.$$

In other words, a (measurable) function ϕ belongs to $\mathbf{X}$ if there exists a positive measure μ on $\mathbf{T}$, with decomposition $d\mu = \varphi\, dt + d\mu_s$, such that $|\phi| \le |\varphi + i\tilde{\mu}|$ a.e. on $\mathbf{T}$. We also define for $\phi \in \mathbf{X}$ the norm

$$\|\phi\|_{\mathbf{X}} = \inf\{\|\mu\| : |\phi| \le |\varphi + i\tilde{\mu}| \text{ a.e.on } \mathbf{T}\}.$$

and pose the following problem:

PROBLEM. *Describe* $\mathbf{X}$ *and* $\|\phi\|_{\mathbf{X}}$ *intrinsically.*

For the time being we note that trivially $L^1 \subset \mathbf{X}$ and in the other direction, by Kolmogorov's theorem, $\mathbf{X}$ is contained in $L^{1,\infty}$, weak-L^1 . The problem of describing $\mathbf{X}$ is the problem of describing the size of the functions in Σ.

All these comes from [6] and [2]. The motivation from this problem comes from an open question in complex function theory in the unit disc, the characterization of the peak sets for the class Λ_α of holomorphic functions in $\mathbf{D}$ satisfying

$$|f(z) - f(w)| = O(|z-w|^\alpha), z, w \in \mathbf{D}, 0 < \alpha \le 1.$$

(A closed set $\mathbf{E} \subset \mathbf{T}$ is called a *peak set* for Λ_α if it exists $f \in \Lambda_\alpha$ such that $f = 1$ on E and $|f| < 1$ on $\overline{\mathbf{D}} \setminus E$.) Using the transformations $f \to e^{-f}$ and $f \to 1 - f$ it is immediate to see that $\mathbf{E}$ is a peak set for Λ_α if there exists $g \in \Lambda_\alpha$ such that $g = 0$ on E and $Reg > 0$ on $\overline{\mathbf{D}} \setminus E$. Then $|g(z)| = O(d_E^\alpha(z))$, where $d_E(z) =$ distance from z to E, and since $g^{-1} \in \Sigma$, it follows that $d_E^{-\alpha} \in \mathbf{X}$. It seems then reasonable to try to understand first $\mathbf{X}$. Incidentally, we point out that the condition $d_E^{-\alpha} \in L^{1,\infty}$ implies when $\alpha = 1$ that E is a finite set, and so the question is interesting only in case $0 < \alpha < 1$.

Among the functional analysis properties of $\mathbf{X}$ we quote the following result of Noell and Wolff:

THEOREM([6]). $\mathbf{X}$ *is a quasi-Banach space.*

This means that the norm $\|.\|$ satisfies

$$\|\phi_1 + \phi_2\|_{\mathbf{X}} \leq c\,(\|\phi_1\|_{\mathbf{X}} + \|\phi_2\|_{\mathbf{X}})$$

and that $\mathbf{X}$ is complete with this norm. In [6] it is also proved that if

$$\sum \|\phi_n\|_{\mathbf{X}}(1 + \log^+ \frac{1}{\|\phi_n\|_{\mathbf{X}}}) < \infty,$$

then $\sum \phi_n \in \mathbf{X}$. The proof of the theorem is as follows: let ϕ_1, ϕ_2 be dominated a.e. by $f_1, f_2 \in \Sigma$, respectively, and let μ_1, μ_2 be the measures corresponding to f_1, f_2, so that $f_i(0) = \|\mu_i\|$, i=1,2. The function $f := (f_1^{1/2} + f_2^{1/2})^2$ belongs to Σ and satisfies

$$|f(z)| \geq |f_1(z)| + |f_2(z)| \geq |\phi_1| + |\phi_2| \geq |\phi_1 + \phi_2|,$$

thus proving that $\phi_1 + \phi_2 \in \mathbf{X}$ and

$$\|\phi_1 + \phi_2\|_{\mathbf{X}} \leq |f(0)| = (\|\mu_1\|^{1/2} + \|\mu_2\|^{1/2})^2 \leq 2(\|\mu_1\| + \|\mu_2\|).$$

Since $\|\mu_i\|$ is arbitrarily close to $\|\phi_i\|_{\mathbf{X}}$, we are done.

One consequence of the above is that one can use signed measures in the definition of $\mathbf{X}$ without enlarging the class.

The problem stated is related to a question raised by A. Baernstein concerning the class Σ. It is easy to see ([4,pg 279]) that if $f \in \Sigma$ then $w = \log|f|$ satisfies

$$|\{t \in I : |w(e^{it} - w(I)| > \delta\}| \leq c|I|e^{-\delta}, \delta > 0 \tag{1}$$

for all intervals I, where $w(I)$ denotes the mean of w over I and c does not depend on I. Some time ago, A.Baernstein conjectured that a positive function ϕ on $\mathbf{T}$ has the same order of magnitude as a function in Σ, or which is the same, $\log\phi = u + \tilde{v}$ with $\|v\|_\infty \leq \pi/2$, if and only if $w = \log\phi$ satisfies (1). This condition appears as the limiting case of

$$|\{t \in I : |w(t) - w(I)| > \delta\}| \leq c|I|e^{-p\delta}, \delta > 0$$

for some $p > 1$, describing those ϕ for which $\log\phi = u + \tilde{v}$ with $\|v\|_\infty < \pi/2$, or which is the same, ϕ is equivalent to $|f|$ for some holomorphic f taking values in a sector of opening strictly less than π (these are also the A_2 weights, see [4]). Baernstein's conjecture was disproved by Wolff and in fact another counterexample is given in [6].

Since for the class $\mathbf{X}$ what are relevant are the size properties of functions in $\mathbf{\Sigma}$ rather than the more subtle equilibrium conditions like (1), one could think that the problem stated should be in principle easier than the question of describing those ϕ with $\log\phi = u + \tilde{v}, \|v\|_\infty \leq \pi/2$.

In the following some properties of functions in $\mathbf{\Sigma}$ will be described. Some of these properties will be "size" properties and will give information on $\mathbf{X}$. Others will be of "equilibrium" type and these will lead to some new notions.

2. The weak maximal function. One way of saying something more about $\mathbf{X}$ is to try to strengthen Kolmogorov's theorem. This is the statement that

$$\delta|\{t : |f(t)| > \delta\}| \leq C|f(0)|, f \in \mathbf{\Sigma}, \delta > 0,$$

i.e. $\|f\|_{L^{1,\infty}(T)} \leq c|f(0)|$ (here and in the following, given a measure μ, $\|\varphi\|_{L^{1,\infty}(\mu)}$ will denote the "weak L_1 norm" sup $\delta\,\mu\{t : |\varphi(t)| > \delta\}$). The conformal invariant form of Kolmogorov's theorem is

$$\|f\|_{L^{1,\infty}(\omega_z)} \leq C|f(z)|, f \in \mathbf{\Sigma}, z \in \mathbf{D},$$

where $d\omega_z(t) = 1 - |z|^2/|z - e^{it}|^2 dt$ is the harmonic measure of the point z. Applying this to f^{-1} and using the fact that

$$\|\varphi\|_{L^{1,\infty}(\mu)}\|\varphi^{-1}\|_{L^{1,\infty}(\mu)} \geq 1/4, \|\mu\| = 1,$$

(weak Jensen's inequality) one obtains that

$$|f(z)| \simeq \|f\|_{L^{1,\infty}(\omega_z)}, f \in \mathbf{\Sigma}, z \in \mathbf{D}. \tag{2}$$

The estimate $|f(z)| \geq c\,\|f\|_{L^{1,\infty}(\omega_z)}$ can be deduced from (1).

The relation (2) shows that $|f|$ in $\mathbf{D}$ can be essentially recovered from its boundary values. We remark in passing that (2) can be seen as the limiting case of the following fact: if an holomorphic function f takes values in a sector of opening α less than π then $|f(z)| \simeq \|f\|_{L^1(\omega_z)}$, (but with constants that become infinite as α approaches π).

Obviously (2) implies

$$\|f\|_{L^{1,\infty}(\omega_z)}\|f^{-1}\|_{L^{1,\infty}(\omega_z)} \leq C^2, f \in \mathbf{\Sigma},$$

and from this it easily follows that for all intervals I

$$\|f\|_{L^{1,\infty}(\chi_I dt)}\|f^{-1}\|_{L^{1,\infty}(\chi_I dt)} \leq C^2|I|^2, f \in \mathbf{\Sigma}.$$

This could be called the "weak-A_2-condition", as it can be thought, again, as the limiting case of the A_2-condition satisfied by $|f|$ when f takes values in a sector of opening less than π.

Now we will see how the estimate (2) leads in a natural way to the notion of the weak maximal function. If f^* denotes the radial maximal function $f^*(t) = \sup_r |f(re^{it})|$, we obtain from (2)

$$|f^*(t)| \simeq \sup_r \|f\|_{L^{1,\infty}(\omega_{re^{it}})}.$$

Now it is easy to see that the measures $\omega_{re^{it}}$, $0 \leq r < 1$ can be replaced by the measures $\frac{1}{|I|}\chi_I dt$, I centered at e^{it}, i.e.

$$(3) \qquad f^*(t) \simeq \sup_I \frac{1}{|I|} \|f\|_{L^{1,\infty}(\chi_I dt)} \simeq \sup_I \frac{1}{|I|} \sup_\delta \delta\, |\{x \in I : |f(e^{ix})| > \delta\}|.$$

Thus we have arrived to the following definition:

DEFINITION. *Given $\phi \in L^{1,\infty}$, its weak maximal function $M_w\phi$ is defined at t by*

$$M_w\phi(t) = \sup_I \frac{1}{|I|} \sup_\delta \delta |\{x \in I : |\phi(e^{ix})| > \delta\}|,$$

the supremum being taken over all the intervals I centered at t.

Observe that the definition is analogous to the one of the Hardy-Littlewood maximal function but replacing the L^1-norm on I by the "weak L^1-norm" on I. Hence (3) reads

$$(4) \qquad f^* \simeq M_w f, f \in \mathbf{\Sigma}$$

suggesting that $M_w f$ is a real-variable version of f^*. It is easily checked that $M_w\phi(t)$ can be alternatively defined

$$M_w\phi(t) = \sup_\delta M\phi_\delta(t)$$

where ϕ_δ is the truncation

$$\phi_\delta(t) = \begin{cases} \delta \text{ if } |\phi(t)| > \delta \\ 0 \text{ otherwise} \end{cases}$$

and $M\phi_\delta$ its Hardy-Littlewood maximal function. Also, $M_w\phi \geq \phi$ a.e.

It should be pointed out that previously E. M. Stein had considered this kind of maximal function, even for general Lorentz spaces $L^{p,q}$, in [10]. In another setting it is showed there that the analogue of M_w in $L^{p,q}$ has some nice properties if $q \leq p$ but not for $q > p$.

Still one more interpretation of M_w can be given as follows. For a weak-L^1 positive function ϕ on $\mathbf{T}$ let us define

$$\tilde{P}[\phi](z) = \|\phi\|_{L^{1,\infty}(\omega_z)} = \sup_\delta \delta\, \omega_z\{t : \phi(e^{it}) > \delta\}.$$

Then $\tilde{P}[\phi]$ is a subharmonic function in the disc $\mathbf{D}$ taking the given boundary values $\phi(e^{it})$. The above shows that

$$M_w\phi \simeq \tilde{P}[\phi]^*,$$

the radial maximal function of $\tilde{P}[\phi]$. This is to be compared with $M\phi \simeq P[\phi]^*$, where $\phi \in L^1, M\phi$ is its Hardy-Littlewood maximal function and $P[\phi]$ its Poisson transform.

Up to now we have thus introduced M_w and we have seen that for $f \in \mathbf{\Sigma}$ one has $f^* \simeq M_w f$. But, as it is well known([4,pg 111]), f^* is again in $L^{1,\infty}$ so we could apply M_w once again. It turns out that the following holds:

THEOREM([2]). *There exists a constant C such that $M_w f^* \leq Cf^*$ for all $f \in \Sigma$.*

Now it is clear that the above theorem, (4) and the monotonicity of M_w together imply the following corollary:

COROLLARY. *There exists a constant C such that for every $\phi \in \mathbf{X}$,*

$$\sup_m C^m M_w^m \phi \in L^{1,\infty}. \tag{5}$$

Here M_w^m denotes the m-th iteration of M_w. Thus, not only every $\phi \in \mathbf{X}$ is in $L^{1,\infty}$ but all its "orbit" under M_w remains in $L^{1,\infty}$ and their weak-L^1 norms growth exponentially. We point out that this property is really relevant in our problem, in the sense that it is not shared by an arbitrary $\phi \in L^{1,\infty}$. Indeed, given m, a function $\phi \in L^{1,\infty}$ can be constructed such that $M_w\phi, \ldots, M_w^{m-1}\phi$ are in $L^{1,\infty}$ but $M_w^m\phi$ is not.

3. An analogous problem for the Hardy-Littlewood maximal function. Given the analogous boundedness properties of the conjugate operator and the Hardy-Little wood maximal operator, it is natural to introduce as well the class

$$\mathbf{Y} = \{\phi : \mathbf{T} \longrightarrow \mathbf{R};\ \exists \mu \geq 0 \text{ s.t. } |\phi| \leq M\mu \text{ a.e.on } \mathbf{T}\}$$

and for $\phi \in \mathbf{Y}$ the norm

$$\|\phi\|_{\mathbf{Y}} = \inf\{\|\mu\| : |\phi| \leq M\mu \text{ a.e.}\}$$

and state as well the problem of describing $\mathbf{Y}$ and $\|\phi\|_{\mathbf{Y}}$ intrinsically.

In this case it is obvious that $\mathbf{Y}$ is a quasi-Banach space and, again, $L^1 \subset \mathbf{Y} \subset L^{1,\infty}$. It turns out that a similar development can be done here. The statement (4) for all $f \in \Sigma$ has as analogue the statement

$$M_w M\mu \simeq M\mu \tag{6}$$

for all measures $\mu \geq 0$. In fact, an n-dimensional version of this is true (a proof is implicit in [3]). Consequently, the corollary of the previous section also holds with $\mathbf{X}$ replaced by $\mathbf{Y}$.

Relations (4) and (6) could be restated by saying that f^* and $M\mu$ are "weak-A_1 -weights". The relationship with the usual A_1-weights is not just formal since using

$$M\psi^p \leq \frac{1}{1-p}(M_w\psi)^p, p < 1$$

it is immediately seen that ψ^p is an A_1-weight if ψ is a weak-A_1-weight, $p < 1$. In particular this is true for $M\mu$ (this is well known, see [4]) and also for $f^*, f \in \Sigma$.

We now make the trivial observation that if in the definition of $\mathbf{X}$ and $\mathbf{Y}$ one replaces "measures" by "functions in $L^p, p > 1$" then the answer to both problems would be simply $\phi \in L^p$. This is because of the L^p-boundedness of both the conjugate operator and the Hardy-Littlewood maximal operator for $p > 1$. Thus in some sense $\mathbf{X}$ and $\mathbf{Y}$

would be substitutes for L^1. Now, it naturally arises the question of whether $\mathbf{X} = \mathbf{Y}$, asked in [2]. It turns out that neither $\mathbf{X} \subset \mathbf{Y}$ nor $\mathbf{Y} \subset \mathbf{X}$. Namely, in [6] it is showed that if E is the Cantor 1/3-set and μ is the Cantor-measure then $M\mu \notin \mathbf{X}$ and, in the other direction, an integrable function whose Hilbert transform cannot be essentially dominated by any $M\mu$ is shown to exists in [7].

A related problem has been considered in [8].

4. The localization of the weak (1,1)-estimate. The fact that $\mathbf{X} \neq \mathbf{Y}$ already shows that condition (5) cannot be sufficient for neither $\phi \in \mathbf{X}$ nor $\phi \in \mathbf{Y}$. In the paper [6], A.Noell and T.Wolff have showed that the functions in $\mathbf{X}$ and $\mathbf{Y}$ must satisfy other type of conditions. To state their result we need a definition:

DEFINITION. *If $\eta > 1$ and N is an integer, a family $\mathcal{F}$ of intervals is called (η, N)-disjoint if any point belongs at most to N intervals ηI, with $I \in \mathcal{F}$ (here ηI denotes the interval having the same center as I and length $\eta|I|$).*

THEOREM ([6]). *There exists a constant C_o such that whenever $\mathcal{F}$ is a (η, N)-disjoint family of intervals, then for all measures μ*

$$\sum_{I\in\mathcal{F}} \sup_{\delta\geq c\|\mu\|/|I|} \delta|\{x \in I : |\tilde{\mu}(x)| > \delta\}| \leq C_o N\|\mu\| \tag{7}$$

for some constant $c = c(\eta)$. The same statement holds for $M\mu$.

The proof is very simple and it can be given here. If I is an interval, set $\mu_1 = \chi_{\eta I}\mu$ and $\mu_2 = \mu - \mu_1$.For $x \in I$,

$$|\tilde{\mu}_2(x)| \leq \frac{2\|\mu\|}{(\eta-1)|I|}.$$

Hence if $\delta > 4\|\mu\|/(\eta-1)|I|$, then since $\tilde{\mu} = \tilde{\mu}_1 + \tilde{\mu}_2$,

$$\{x \in I : |\tilde{\mu}(x)| > \delta\} \subset \{x \in I : |\tilde{\mu}_1(x)| > \delta/2\}.$$

Therefore if C is the constant in the weak (1,1)-estimate

$$|\{x \in I : |\tilde{\mu}(x)| > \delta\}| \leq \frac{2C}{\delta}\|\mu_1\| = \frac{2C}{\delta}\mu(\eta I)$$

and (7) follows adding on $I \in \mathcal{F}$.

Condition (7), which is inherited by functions in $\mathbf{X}$ and $\mathbf{Y}$, is easier to handle than (5) in order to check that a given function is not in $\mathbf{X}$ or $\mathbf{Y}$. For example, if E is the set $E = \{1/n, n = 1, 2, \dots\} \cup \{0\}$ and $\phi = d_E^{-1/2}$ it is quite easy to check that (7) fails for ϕ. On the other hand, this function ϕ satisfies (5) because it can be checked that $M_w\phi \simeq \phi$, thus providing another example showing that (5) is not sufficient for ϕ to be in $\mathbf{X}$ or $\mathbf{Y}$. Also note that since $\mathbf{X} \neq \mathbf{Y}$ condition (7) is not sufficient neither for $\phi \in \mathbf{X}$ nor $\phi \in \mathbf{Y}$.

5. A sharper form of Kolmogorov's theorem. The theorem of Kolmogorov stating that any $f \in \Sigma$ has boundary values in $L^{1,\infty}$ was in the basis of all the development leading to the necessary condition (5). In the previous section we have seen that condition (7) is, as (5), necessary but not sufficient for $\phi \in \mathbf{X}, \mathbf{Y}$. It would be thus interesting to have some condition that takes into account both types of restriction on the growth. A first step in this direction has been recently done by B.Korenblum ([5]) by sharpening Kolmogorov's theorem so as to capture as well condition (7). To describe Korenblum's theorem we begin by modifying a bit the statement of Kolmogorov's theorem.

Introducing the δ-truncation of f by

$$f_\delta(e^{it}) = \begin{matrix} \delta \text{ if } |f(e^{it})| \geq \delta \\ 0 \text{ otherwise} \end{matrix}$$

and if P denotes the Poisson transform , Kolmogorov's theorem is the statement that

$$\sup_\delta P[f_\delta](0) \leq C|f(0)|, f \in \Sigma$$

and

$$\sup_\delta P[f_\delta](z) \leq C|f(z)|$$

its conformally invariant form. B.Korenblum has noticed that one can use other truncations than just truncations with constant functions. Associated to each $z \in \mathbf{D}$ consider the family of functions

$$\mathcal{F}_z = \{\Psi(w) = \lambda \frac{1 - \bar{z}w}{w - \zeta}, |\zeta| = 1, \lambda \in \mathbf{C}\}$$

Note that if τ is a Moebius transformation and $\tau(z) = \eta$, then$\mathcal{F}_\eta \circ \tau = \mathcal{F}_z$. Finally, for a pair of functions f, Ψ the Ψ-truncation of f is defined

$$f_\Psi(z) = \begin{matrix} \Psi(z) \text{ if } |f(z)| \geq |\Psi(z)| \\ 0 \text{ otherwise} \end{matrix}$$

THEOREM ([5]). *If $f \in \Sigma$ then*

$$\sup_{\Psi \in \mathcal{F}_z} \min\{\Psi(z), P[f_\Psi](z)\} \leq C|f(z)|, z \in \mathbf{D}.$$

Korenblum obtains a nice and direct proof of this theorem. It can be proved as well using its conformal invariance and (7).

6. A final remark on Calderón-Zygmund singular convolution operators. We have obtained in (5) the statement

$$\sup_m C^m M_w^m \tilde{\mu} \in L^{1,\infty}$$

and the proof has been based on complex function theory, namely properties of the class Σ. It is natural to ask whether a real analysis proof of this can be given. This turns out to be the case , and even with the conjugate operator replaced by any Calderón-Zygmund singular convolution operator.

In [1] it is showed that if T is a singular convolution operator associated to a kernel K satisfying

$$|K(x)| \leq C|x|^{-n}$$
$$|\nabla K(x)| \leq C|x|^{-n-1}$$

and T^* its corresponding maximal operator (see [9] for definitions) then

$$M_w T^* f \leq C(T^* f + Mf), f \in L^1.$$

Obviously, this together with (6) implies that

$$\sup_m C^m M_w^m Tf \in L^{1,\infty}$$

whenever T^* satisfies a weak $(1,1)$-estimate.

References

1. J.Bruna and B.Korenblum, *A note on Calderón-Zygmund singular integral convolution operators*, Bull.Amer.Math.Soc. **16(2)** (April 1987).
2. J.Bruna and B.Korenblum, *On Kolmogorov's theorem, the Hardy-littlewood maximal function and the radial maximal function*, J.d'Analyse Mathématique **50**, 225–239, (1988).
3. R.R.Coifman and R.Rochberg, *Another characterization of B.M.O.*, Proc. Amer. Math. Soc. **79**, 1980.
4. J.Garnett, "Bounded analytic functions" Academic Press. New York.
5. B.Korenblum, *A sharper form of a theorem of Kolmogorov*, preprint.
6. A.Noell and T.Wolff, *Peak sets for Lip α classes*, preprint.
7. A.Samotij, *An example of a Hilbert transform*, preprint.
8. P.Sjögren, *How to recognize a discrete maximal function*, preprint.
9. E.M.Stein, "Singular integrals and differentiability properties of functions", Princeton Univ.Press. Princeton.
10. E.M.Stein, *Editor's note: The differentiability of functions in* $\mathbf{R}^n$, Annals of Math., **133(2)**, (1981).

ω-CALDERÓN-ZYGMUND OPERATORS

R. COIFMAN, G. DAVID, Y. MEYER, AND S. SEMMES

Abstract. We enlarge the class of singular integrals of Calderón-Zygmund type by generalizing the usual assumptions on the kernel. These weaker conditions on the kernel arise naturally in the study of operators depending (linearly or not) on a functional parameter. Examples include the Cauchy integral operator, viewed as a function of the curve, and multilinear operators, viewed as operating on one of the arguments while the others are frozen.

§1. INTRODUCTION

Let T be a linear operator mapping test functions on $\mathbf{R}^n$ continuously into distributions and with an associated kernel $K(x,y)$, $x \neq y$ (in the sense that $< Tf, g >= \int\int g(x)K(x,y)f(y)dxdy$ whenever f and g are test functions with disjoint support). Let w be an A_1 weight, and set $\omega_t(x) = t^{-n}\int_{|x-y|<t} \omega(y)dy$. We say that $K(x,y)$ satisfies ω-standard estimates if for some $\varepsilon > 0$,

$$|K(x,y)| \leq C\omega_{|x-y|}(x)\frac{1}{|x-y|^n} \tag{1.1}$$

and

(1.2)

$$|K(x,y) - K(x',y)| \leq C\omega_{|x-x'|}(x)\frac{|x'-x|^\varepsilon}{|x-y|^{n+\varepsilon}} \quad when \quad |x-x'| < \frac{1}{2}|x-y|$$

$$|K(x,y) - K(x,y')| \leq C\omega_{|y-y'|}(y)\frac{|y-y'|^\varepsilon}{|x-y|^{n+\varepsilon}} \quad when \quad |y-y'| \leq \frac{1}{2}|x-y|.$$

The simplest example of such an operator is the Calderón commutator

$$C(a,f) = P.V.\int_{-\infty}^{\infty} \frac{A(x)-A(y)}{(x-y)^2} f(y)dy, A' = a, \tag{1.3}$$

where we can take $\omega = ((a^{\frac{1}{\delta}})^*)^\delta$ for any δ, $0 < \delta < 1$. More generally, we can consider bilinear (or multilinear) singular integrals, as in [CM1].

Another example is given by the Cauchy integral. Let A be a real-valued locally absolutely continuous function and define

$$T_A f(x) = P.V.\int_{-\infty}^{\infty} \frac{1}{x-y+i(A(x)-A(y))} f(y)dy. \tag{1.4}$$

The kernel of T_A satisfies (1.1) with $\omega \equiv 1$ and (1.2) with $\omega = ((A')^*)^\varepsilon$ for $0 \le \varepsilon \le 1$.

In §2 we prove a version of the $T(1)$ theorem. That is, we prove that an operator T with ω-standard kernel maps $L^2(\omega)$ boundedly into $L^2(\omega^{-1})$ if T satisfies a weak boundedness property and if $T1$ and T^t1 lie in BMO_ω. (BMO_ω can be defined as the space of locally integrable functions f such that $f^\sharp \le C\omega$.)

A consequence of this is that if T is associated to a 1- standard kernel and satisfies the weak boundedness property, then T maps $L^2(\omega)$ into $L^2(\omega^{-1})$ with $\omega = 1 + S_r(T(1)) + S_r(T^t(1))$. Here

$$S_r(f)(x) = \sup_{Q \ni x}(|Q|^{-1} \int_Q |f(y) - m_Q f|^r dy)^{1/r}, m_Q f = \frac{1}{|Q|}\int_Q f,$$

and any $r > 1$ is allowed in this choice of ω. Results of this type have been obtained by Leckband [L]. He showed that one could get weighted estimates for an operator T associated to a standard kernel but for which $T1$ and T^t1 do not lie in BMO but are only locally in L^p for some $p > 2$.

As another application, we use the $T(1)$ theorem to indicate a new proof that for chord-arc curves with small constant the difference between the Cauchy integral and the Hilbert transform is small [CM2]. The usual $T(1)$ theorem doesn't seem to apply here, since this difference is not a Calderón-Zygmund operator.

Let us call a linear operator T an ω-Calderón-Zygmund operator ($\omega - CZO$) if it has a kernel satisfying the ω- standard estimates (1.1) and (1.2) and if T maps $L^2(\omega)$ boundedly into $L^2(\omega^{-1})$. The operators (1.3) and (1.4) are examples of this. For (1.3) we take $\omega = (a^{\frac{1}{\delta}})^{*\delta}, 0 < \delta < 1$, while for (1.4) we take $\varepsilon = \frac{1}{2}$, $\omega = (A')^{*\frac{1}{2}}$. The choice of $\varepsilon = \frac{1}{2}$ comes from Murai's estimates for the Cauchy integral.

In §3 we develop a version of Calderón-Zygmund theory for $\omega - CZO's$. We show for example that an $\omega - CZO$ satisfies a full range of L^p estimates, as well as a weak-type result for $p = 1$ and an L^∞ into BMO_ω result for $p = \infty$. Many of the proof details are the same as in the classical case and hence are omitted and left to the reader. However, the ingredients have to be set up differently than usual, and it is these differences that will be emphasized.

The authors were partially supported by the NSF.

§2. A $T1$ theorem for ω-CZO's.

Let T be a continuous linear operator from test functions to distributions. We say that T satisfies the $\omega - WBP$ (weak boundedness property) if for each ball B and any test functions ϕ_1, ϕ_2 supported in B that satisfy $\|\phi_i'\|_\infty \leq |B|^{-\frac{1}{n}}$ we have

$$| < T\phi_1, \phi_2 > | \leq C\omega(B).$$

This will hold if T is an $\omega - CZO$, or if T is associated to an antisymmetric kernel satisfying the first standard estimate (1.6).

THEOREM 2.1.. *Let T be a continuous linear operator from test functions to distributions that is associated to an ω- standard kernel. Then T is an $\omega - CZO$ if and only if T satisfies the $\omega - WBP$ and if $T1$ and T^t1 lie in BMO_ω.*

The "only if" part follows from the results of Section 3. Thus we only consider the "if" part. This theorem admits an immediate generalization of itself. Suppose $T1$ or T^t1 don't lie in BMO_ω. Let $S_r(f)$ be as in the Introduction. For $r > 1$, $S_r(f)$ is an A_1 weight with constant depending only on r, unless $S_r(f) \equiv \infty$. This is completely analogous to the corresponding result for the maximal function. (See [J], p. 19.) Thus if $T1$ or T^t1 don't lie in BMO_ω, and if $S_r(T1) \not\equiv \infty$ and $S_r(T^t1) \not\equiv \infty$ for some $r > 1$, then we can replace ω by $\tilde{\omega} = \omega + S_r(T1) + S_r(T^t1)$. Then the kernel of T is $\tilde{\omega}$-standard, and now $T1$ and T^t1 lie in $BMO_{\tilde{\omega}}$. If T satisfies the $\omega - WBP$, then we can conclude that T is an $\tilde{\omega} - CZO$.

Before discussing the proof of the theorem, let us make some observations concerning BMO_ω. First, $f \in BMO_\omega$ if and only if

$$\sup_Q \frac{1}{\omega(Q)} \int_Q |f - m_Q f|^r \omega^{1-r} < \infty \quad for \quad any \quad r \in [1, \infty).$$

Indeed, if this holds for $r = 1$, then it is easy to see that $f^\sharp \leq C\omega^* \leq C\omega$, since $\omega \in A_1$. Conversely, if $f^\sharp \leq C\omega$, then $\int_Q (f^\sharp)^r \omega^{1-r} \leq C\omega(Q)$ for all cubes Q. For any $r < \infty$, $\omega^{1-r} \in A_\infty$, since $\omega \in A_1$. By localizing the good-λ inequality by which, say, the dyadic maximal function is controlled by $f^\sharp$ (see [J], p. 41), we get from the preceeding facts that $\int_Q |f - m_Q f|^r \omega^{1-r} \leq C\omega(Q)$ for dyadic cubes Q, and it is easy to get all cubes from this.

See Garcia-Cuerva [GC] for more about BMO_ω (for $\omega \in A_p$) and connections with weighted Hardy spaces.

Note also that if $f \in BMO_\omega$, then $S_r(f) \leq C\omega$ for some $r > 1$. This can be obtained from the above and the reverse Hölder inequality for ω.

To prove the theorem we must show that T maps $L^2(\omega)$ into $L^2(\frac{1}{\omega})$.

Lemma 2.2. *Suppose that ϕ, ψ are C^∞, supp $\phi, \psi \subseteq \{|x| \leq 1\}$, $\int \psi = 0$, and that $b \in BMO_\omega$. Define $P_t f = \phi_t * f$, $\phi_t(x) = \frac{1}{t^n}\phi(\frac{x}{t})$, and $Q_t f = \psi_t * f$. Then*

$$T_b f = \int_0^\infty Q_t((Q_t b)(P_t f))\frac{dt}{t} \quad \text{is an} \quad \omega - CZO.$$

The proof that the kernel of T_b satisfies ω -standard estimates is fairly straightforward. It is useful to observe that

$$|m_{2^{k+1}Q}b - m_{2^k Q}b| \leq C\|b\|\frac{\omega(2^{k+1}Q)}{|2^{k+1}Q|} \leq C\|b\|\frac{\omega(Q)}{|Q|}$$

so that $|m_{2^k Q}b - m_Q b| \leq Ck\|b\|\frac{\omega(Q)}{|Q|}$.

There are a variety of ways to check that T_b is an $\omega - CZO$. It is enough to show that if $f, g \in L^2(\omega)$, then

$$|\int_0^\infty \int_{\mathbf{R}} Q_t(g)Q_t(b)P_t(f)\frac{dxdt}{t}| \leq C\|f\|_{L^2(\omega)}\|g\|_{L^2(\omega)}\|b\|_{BMO_\omega}.$$

Because $\omega \in A_2$, we have that $\int\int |Q_t g|^2 \omega \frac{dxdt}{t} \leq C\|g\|^2_{L^2(\omega)}$, and we are reduced to

$$\int_0^\infty \int_{\mathbf{R}} |Q_t(b)|^2 |P_t f|^2 \omega^{-1}\frac{dxdt}{t} \leq C\|f\|_{L^2(\omega)}\|b\|_{BMO_\omega}.$$

Again using $\omega \in A_2$ we reduce to showing that we have an ω -Carleson measure, i.e., for any interval I,

$$\int_0^{|I|}\int_I \omega^{-1}|Q_t b|^2\frac{dxdt}{t} \leq C\omega(I).$$

This follows from $\omega^{-1} \in A_2$, the weighted inequalities for the square function $f \mapsto (\int_0^\infty |Q_t f|^2 \frac{dt}{t})^{\frac{1}{2}}$, and the fact that $b \in BMO_\omega$ implies

$$\int_I |b - m_I b|^2 \omega^{-1} dx \leq C\|b\|^2_{BMO_\omega}\omega(I),$$

as observed above.

As usual, Lemma 2.2 allows us to reduce Theorem 2.1 to the case where $T1 = 0$ and $T^t 1 = 0$ (as elements in BMO_ω, i.e., modulo constants). Indeed, if $b_1 = T1$ and $b_2 = T^t 1$, then $R = T - T_{b_1} - T^t_{b_2}$ satisfies the hypotheses of Theorem 2.1 and also $R1 = 0$, $R^t 1 = 0$, at least if we require that $\int \phi = 1$ and $\int_0^\infty Q_t^2 \frac{dt}{t} = I$.

LEMMA 2.3.. *Same notation and hypotheses as in Theorem 2.1 and Lemma 2.2. Assume also that $T1 = 0$ and $T^t 1 = 0$. Then the kernel $K_{s,t}(x, y)$ of $Q_s T Q_t$ satisfies:*

$$(i) \quad \text{for} \quad 0 \le s \le t, |K_{s,t}(x,y)| \le (\frac{s}{t})^\varepsilon \frac{t^\varepsilon}{(|x-y|+t)^{n+\varepsilon}} \omega_s(x);$$

and

$$(ii) \quad \text{for} \quad 0 \le t \le s, |K_{s,t}(x,y)| \le (\frac{t}{s})^\varepsilon \frac{s^\varepsilon}{(|x-y|)^{n+\varepsilon}} \omega_t(x).$$

It suffices to prove (i). Formally, we have

$$K_{s,t}(x,y) = \int\int \psi_s(x-v) K(v,u) \psi_t(u-y) du dv. \tag{2.4}$$

Because supp $\psi \subseteq \{|x| < 1\}$, we have that $|x - v| < s$ and $|u - y| < t$, in addition to the assumption that $s < t$. Assume first that $|x - y| \ge 4t$, so that $|(x - v) - (y - u)| > 2t$. Then $\psi_s(x - \cdot)$ and $\psi_t(\cdot - y)$ have disjoint supports, and (2.4) is valid. Using $\int \psi_s = 0$ we get

$$|K_{s,t}(x,y)| = |\int\int \psi_s(x-v)\{K(v,u) - K(x,u)\}\psi_t(u-y) du dv|$$

by (1.2)

$$\le C \int\int |\psi_s(x-v)| \frac{|x-v|^\varepsilon \omega_s(x)}{|x-y|^{n+\varepsilon}} |\psi_t(u-y)| du dv \le C \frac{s^\varepsilon}{(|x-y|+t)^{n+\varepsilon}} \omega_s(x).$$

Now assume that $|x - y| \le 4t$. We shall calculate formally, using (2.4), but one can easily make the argument rigorous, just as with the usual $T(1)$ theorem. We can write $1 = \eta_0(\frac{|x-y|}{s}) + \eta_1(\frac{|x-y|}{s})$, where $\eta_0, \eta_1 \in C^\infty$, supp $\eta_0 \subseteq [-3, 3], \eta_0 \equiv 1$ on $[-2, 2]$. Then

$$\begin{aligned} K_{s,t}(x,y) &= \int\int \psi_s(x-v) K(v,u)[\psi_t(u-y) - \psi_t(x-y)] du dv \quad (\textit{since} \quad T^t 1 = 0) \\ &= \int\int \psi_s(x-v) K(v,u)[\psi_t(u-y) - \psi_t(x-y)] \eta_0\left(\frac{|x-u|}{s}\right) du dv \\ &+ \int\int \psi_s(x-v) K(v,u)[\psi_t(u-y) - \psi_t(x-y)] \eta_1\left(\frac{|x-u|}{s}\right) du dv \\ &= A + B. \end{aligned}$$

For the first term we use the $\omega - WBP$ applied to the ball $B = \{u : |x - u| \le 2s\}$, with $\phi_1(v) = s^n \psi_s(x - v)$ and

$$\varphi_2(u) = t^{n+1}s^{-1}[\psi_t(u-y) - \psi_t(x-y)]\eta_0(\frac{|x-y|}{s}).$$

Then supp $\varphi_1, \varphi_2 \subseteq B$ and $\|\phi_i\|_\infty \le C|B|^{-\frac{1}{n}}$, and the $\omega - WBP$ gives

$$|A| \le Cs^{-n}t^{-(n+1)}s\omega(B) \le C\frac{s}{t^{n+1}}\omega_s(x).$$

For the second term we use the second standard estimate (1.2); because $\int \psi_s = 0$, we can replace $K(v,u)$ by $K(v,u) - K(x,u)$ to get

$$\begin{aligned}|B| &\le C\int\int |\psi_s(x-v)|\frac{|x-v|^\varepsilon \omega_s(x)}{|x-u|^{n+\varepsilon}}|\psi_t(u-y) - \psi_t(x-y)|\eta_1(\frac{|x-u|}{s})dudv \\ &\le C\int\int_{2s\le|x-u|\le t} |\psi_s(x-v)|\frac{s^\varepsilon\omega_s(x)}{|x-u|^{n+\varepsilon}}\frac{|x-u|}{t^{n+1}}dudv \\ &+ C\int\int_{t\le|x-u|<\infty} |\psi_s(x-v)|\frac{s^\varepsilon\omega_s(x)}{|x-u|^{n+\varepsilon}}\frac{1}{t^n}dudv \\ &\le Cs^\varepsilon t^{-n-\varepsilon}\omega_s(x) + Cs^\varepsilon t^{-n-\varepsilon}\omega_s(x).\end{aligned}$$

This proves Lemma 2.3.

To prove Theorem 2.1, we have already seen that it suffices to consider the case $T1 = 0$, $T^t1 = 0$. As in the original $T1$ theorem, we choose ψ so that $\int_0^\infty Q_t^2 \frac{dt}{t} = I$, so that

$$\begin{aligned}T &= \left(\int_0^\infty Q_s^2\frac{ds}{s}\right) T \left(\int_0^\infty Q_t^2\frac{dt}{t}\right) \\ &= \int_0^\infty\int_0^\infty Q_s(Q_sTQ_t)Q_t\frac{dt}{t}\frac{ds}{s}.\end{aligned}$$

The fact that T maps $L^2(\omega)$ into $L^2(\frac{1}{\omega})$ now follows from the next lemma. (The formal manipulations we've made can be justified rigourously just as in the classical case.)

LEMMA 2.5.. *If $K_{s,t}(x,y)$ satisfies the estimates in Lemma 2.3 and $f, g \in L^2(\omega)$, then*

$$\int_0^\infty\int_0^\infty\int_{\mathbf{R}}\int_{\mathbf{R}} |Q_sg(x)||K_{s,t}(x,y)||Q_tf(y)|dxdy\frac{ds}{s}\frac{dt}{t} \le C\|f\|_{L^2(\omega)}\|g\|_{L^2(\omega)}.$$

By assumption, the integral is controlled by

$$\int\int_{s<t}\int_{\mathbf{R}}\int_{\mathbf{R}} |Q_sg(x)|\omega_s(x)(\frac{s}{t})^\varepsilon\frac{t^\varepsilon}{(|x-y|+t)^{n+\varepsilon}}|Q_tf(y)|dxdy\frac{ds}{s}\frac{dt}{t} \tag{2.6}$$

plus an analogous term for $s > t$. Setting $P_t^\epsilon h(x) = \int_{\mathbf{R}^n} \frac{t^\epsilon}{(|x-y|+t)^{n+\epsilon}} h(y)dy$, we rewrite (2.6) and dominate it by

$$C \int_0^\infty \int_{\mathbf{R}^n} \Big(\int_0^t |Q_s g(x)| \big(\frac{s}{t}\big)^\epsilon \frac{ds}{s} \Big) \omega(x) P_t^\epsilon(|Q_t f|)(x) \frac{dxdt}{t}.$$

Apply Cauchy-Schwarz. One factor is

$$\left(\int_0^\infty \int_{\mathbf{R}^n} |P_t^\epsilon(|Q_t f|)(x)|^2 \omega(x) \frac{dxdt}{t} \right)^{\frac{1}{2}} \leq C \left(\int_0^\infty \int_{\mathbf{R}} |Q_t f(x)|^2 \omega(x) \frac{dxdt}{t} \right)^{\frac{1}{2}}$$
$$\leq C\|f\|_{L^2(\omega)}.$$

The first inequality uses the fact that P_t^ϵ is bounded on $L^2(\omega)$ (since $\omega \in A_2$), while the second uses the boundedness of $f \mapsto (\int_0^\infty |Q_t f|^2 \frac{dt}{t})^{\frac{1}{2}}$ on $L^2(\omega)$.

The remaining factor is dominated (using Jensen's inequality) by

$$C \int_0^\infty \int_{\mathbf{R}^n} \int_0^t |Q_s g(x)|^2 \big(\frac{s}{t}\big)^\epsilon \frac{ds}{s} \omega(x) \frac{dxdt}{t} \leq C \int_0^\infty \int_{\mathbf{R}^n} |Q_s g(x)|^2 \omega(x) \frac{dxds}{s}$$
$$\leq C\|g\|_{L^2(\omega)}.$$

This completes the proof of Lemma 2.5, and hence also the proof of Theorem 2.1.

Let us give an application of Theorem 2.1 (as mentioned in the Introduction). Let Γ be a rectifiable Jordan curve in the plane. We assume that Γ is a chord-arc curve with small constant. This means that if $z(\cdot)$ is an arclength parameterization of Γ, then there is a small $k > 0$ so that $|s-t| \leq (1+k)|z(s)-z(t)|$ for all $s, t \in \mathbf{R}$. A short calculation shows that this is equivalent to z' having small BMO norm, with $k \approx \|z'\|_*^2$. (See [CM2].)

We want to show that the operator

$$f \mapsto PV \int_{-\infty}^\infty \left(\frac{z'(y)}{z(x)-z(y)} - \frac{1}{x-y} \right) f(y)dy = (C_\Gamma - H)f$$

has small norm if k is small. One can show that C_Γ has norm $\leq C$ using the classical $T1$ theorem. Let us indicate how one can prove that $\|C_\Gamma - H\|_{2,2} \leq C\|z'\|_*$ if k is small. We shall only hit the main points and leave the details to the reader.

It is enough to show that $f \mapsto \chi_I(C_\Gamma - H)(\chi_I f)$ maps $L^{\frac{3}{2}}(|I|^{-1}dx)$ into $L^1(|I|^{-1}dx)$ with norm $C\|z'\|_*$ for all intervals I. Indeed, one can then globalize using good-λ or the sharp function, exactly as in the proofs that $[b, T]$ is bounded if $b \in BMO$ and T is a CZO. (See [CRW],[Ja].)

This can be further reduced as follows. For any interval I, let ϕ_I be a bump such that $0 \le \phi_I \le 1$, $\phi_I = 1$ on I, $\phi_I = 0$ off $2I$, and $|\phi|_\infty \le C|I|^{-1}$. Define R_I by

$$R_I f(x) = \phi_I(x) PV \int \left(\frac{m_I(z')}{z(x) - z(y)} - \frac{1}{x - y} \right) \phi_I(y) f(y) dy.$$

It suffices to show that R_I maps $L^{\frac{3}{2}}(|I|^{-1}dx)$ into $L^1(|I|^{-1}dx)$ with norm $\le C\|z'\|_*$. Indeed, to replace $m_I(z')$ by z', the error term produced is controlled using the boundedness of C_Γ and BMO condition on z'.

The kernel of R_I satisfies ω -standard estimates, with $\omega = (((z' - m_I z')\chi_I)^2)^{*\frac{1}{2}}$. Also, it is not difficult to estimate $R_I(1)$. The idea is to first replace $\phi_I(y)$ by $\phi_I(x)$, and then replace $m_I(z')$ by z'. Both steps give rise to controllable errors. The remaining piece can be computed directly.

§3. Calderón-Zygmund theory for ω-CZO's.

Before plunging into details let us think a moment about what we are looking for.

First of all, what sort of estimates are natural to expect for an ω-CZO? The trivial example that one should keep in mind is multiplication by ω. Thus we can hope that an ω-CZO will map $L^p(\omega^p)$ into L^p and send L^p into $L^p(\omega^{-p})$ for $1 < p < \infty$. By interpolation one expects results in between, such as $L^2(\omega) \to L^2(\omega^{-1})$. An endpoint result like $L^1(\omega) \to$ weak L^1 would not be unwarranted.

What should be the correct maximal function for controlling ω-CZO's? This turns out to be tricky. The first guess might be

(3.1) $$f \mapsto \omega^* f^* \approx \omega f^*.$$

This takes $L^p \to L^p(\omega^{-p})$ for $1 < p < \infty$, but it takes $L^p(\omega^p) \to L^p$ only for p near 1. However, an estimate like $L^p \to L^p(\omega^{-p})$ for ω-CZO's implies $L^p(\omega^p) \to L^p$ by duality. Thus we would expect ω- CZO's to be controlled by a maximal function that has both kinds of estimates.

A slightly better maximal function is

(3.2) $$f \longmapsto \sup_{t>0} \omega_t(x) P_t^\varepsilon f(x),$$

where $P_t^\varepsilon f(x) = \int_{\mathbf{R}^n} |f(u)| \frac{t^\varepsilon}{(|x-u|+t)^{n+\varepsilon}} du$, and $\omega_t(x)$ is as in the Introduction (just before (1.1)). This is pointwise dominated by (3.1), and maps $L^p \to L^p(\omega^{-p})$ if $1 < p < p_\varepsilon$, where $p_\varepsilon > 1$ and $p_1 = \infty$ when $n = 1$. (This last is obtained from the reverse Hölder inequality for ω.) Even if $\varepsilon < 1$, we might be able to improve the range of p if we knew more about ω (as in the case of the Cauchy integral (1.4)).

The operation (3.2) is still not satisfactory, since the $L^p(\omega^p) \to L^p$ estimate will not work in general for $1 < p < \infty$, while one would get such an estimate for ω -CZO's by duality. In terms of the standard estimates (1.1) and (1.2), the difficulty is that while you win decay at infinity in (1.2), for some weights you may loose from replacing $\omega_{|x-y|}(x)$ in (1.1) with $\omega_{|x-x'|}(x)$ in (1.2). To address this we observe that (1.1) and (1.2) imply that for any $\beta \in [0,1]$,

$$|K(x,y) - K(x',y)| \le C(\omega_{|x-y|}(x))^{1-\beta}(\omega_{|x-x'|}(x))^{\beta}\frac{|x-x'|^{\beta\varepsilon}}{|x-y|^{n+\beta\varepsilon}} \tag{3.3}$$

when $|x-x'| \le \frac{1}{2}|x-y|$, and similarly for x and y interchanged. The maximal function naturally associated to standard estimates like (3.3) is

$$\lambda_\omega^{\beta,\varepsilon}(f)(x) = \sup_{t>0}\sum_{i=0}^{\infty} 2^{-j\beta\varepsilon}\omega_t(x)^{\beta}\omega_{2^jt}(x)^{1-\beta}\frac{1}{(2^jt)^n}\int_{|x-u|<2^jt}|f|du \tag{3.4}$$

This is pointwise controlled by (3.2) with ε replaced by $\beta\varepsilon$. In particular, it maps L^p into $L^p(\omega^{-p})$. On the other hand, if $\beta, \varepsilon > 0$, $\beta p \le 1$, and $p > 1$, then $\lambda_\omega^{\beta,\varepsilon}$ maps $L^p(\omega^p)$ into L^p. Indeed, by the A_1 condition on ω,

$$\lambda_\omega^{\beta,\varepsilon}(f) \le C\omega(x)^{\beta}\sup_{t>0} P_t^{\beta\varepsilon}(\omega^{(1-\beta)}f) \le C\omega^{\beta}(\omega^{1-\beta}f)^*), \tag{3.5}$$

and this last operation takes $f \in L^p(\omega^p)$ into L^p if $\omega^{\beta p} \in A_1$. In the context of ω -CZO's, so that we have (3.3), we will be allowed to take any $\beta \in (0,1]$, and ε will be positive.

Ideally, the sort of maximal function we'd like to use is

$$f \longmapsto sup_{t>0}\omega_t(x)\frac{1}{t^n}\int_{|x-u|<t}|f|du. \tag{3.6}$$

If we could control ω -CZO's in terms of this, that would be optimal in a reasonable sense: if we smooth up the average of f, (3.6) can be viewed as a vector-valued ω -CZO, so that estimates for ω -CZO's should hold for it.

The operation $\lambda_\omega^\beta,\varepsilon$ is not so far from (3.6), since we are allowed any β in $(0,1)$, and this is the maximal function we shall use. Although the notation is more cumbersome than for (3.1) and (3.2), it does have the correct L^p estimates.

Now we plunge into details. We first want to show that any reasonable estimate implies an estimate on L^1.

Lemma 3.7. . *Suppose that T is associated to an ω - standard kernel, $\omega \in A_1$, and that for some $p, c > 0$, and all cubes Q we have*

$$(3.8)\qquad \left(\frac{1}{|Q|}\int_Q |Tf|^p dx\right)^{1/p} \le C\omega(Q)\|f\|_\infty \quad if \quad supp \quad f \subseteq Q.$$

Then we also have

$$(3.9)\qquad |Q|^{-1}|\{x \in Q : |Tf| > \lambda\}| \le C\left[\frac{(m_Q\omega)m_Q|f|}{\lambda}\right]^{\frac{p}{p+1}}, \lambda > 0,$$

for all cubes Q and all f supported in Q.

Notice that (3.8 holds if for some $r, C > 0$ and all cubes Q,

(3.8') $\quad T$ maps $L^\infty(Q)$ into $L^r(Q, |Q|^{-1}\omega^{-r}dx)$ with norm $\le C$.

Here one can take the p in (3.8) to be $\frac{r}{(r+1)}$. In particular, (3.8) holds if T is bounded from $L^2(\omega)$ into $L^2(\omega^{-1})$. Conversely, we shall see that the hypotheses of Lemma 3.7 imply that T is an ω -CZO.

To prove the lemma we let $f = g + b$ be a Calderón-Zygmund decomposition of f at height $\gamma\lambda(m_Q\omega)^{-1}$, where $\gamma > 0$ will be chosen later. Thus $\|g\|_\infty \le C\gamma\lambda(m_Q\omega)^{-1}$, so that

$$|Q|^{-1}|\{x \in Q : |Tg| > \frac{\lambda}{2}\}| \le C\lambda^{-p}\int_Q |Tf|^p|Q|^{-1}dx$$

(by (3.8))

$$\le C\lambda^{-p}(\gamma\lambda)^p = C\gamma^p.$$

Also, $b = \sum b_i$, supp $b_i \subseteq Q_i$, $\int b_i = 0$, $m_{Q_i}|b_i| \le C\gamma\lambda(m_Q\omega)^{-1}$. By the ω -standard estimates, if y_i is the center of Q_i, we have

$$\int_{Q\setminus 2Q_i} |Tb_i||Q|^{-1}dx \le C\int_{Q\setminus 2Q_i}\int_{Q_i} m_{Q_i}\omega\frac{|y-y_i|^\epsilon}{|x-y_i|^{n+\epsilon}}|b_i(y)|dy|Q|^{-1}dx$$
$$\le Cm_{Q_i}\omega m_{Q_i}|b_i||Q|^{-1}$$
$$\le C\omega(Q_i)|Q|^{-1}\gamma\lambda(m_Q\omega)^{-1}.$$

Thus, if $E_\lambda = \cup(2Q_i)$, we get

$$|Q|^{-1}|\{x \in Q : |T(b)| > \frac{\lambda}{2}\}| \le |E_\lambda||Q|^{-1} + \lambda^{-1}\int_{Q\setminus E_\lambda} |T(b)||Q|^{-1}dx$$
$$\le C\sum|Q_i||Q|^{-1} + C\sum\omega(Q_i)|Q|^{-1}\gamma(m_Q\omega)^{-1}$$
$$\le C\sum(\int_{Q_i}|f|)(\gamma\lambda)^{-1}m_Q\omega|Q|^{-1} + C\omega(E_\lambda)|Q|^{-1}\gamma(m_Q\omega)^{-1}$$
$$\le C(\gamma\lambda)^{-1}m_Q\omega m_Q|f| + C\gamma.$$

The third inequality follows from $m_{Q_i}|f| > \gamma\lambda(m_Q\omega)^{-1}$, which comes from the definition of the Calderón-Zygmund decomposition.

Altogether, we get

$$\begin{aligned}|Q|^{-1}|\{x \in Q : |Tf| > \lambda\}| &\leq |Q|^{-1}|\{x \in Q : |Tg| > \frac{\lambda}{2}\}| \\ &+ |Q|^{-1}|\{x \in Q : |Tb| > \frac{\lambda}{2}\}| \\ &\leq C\gamma^p + C\gamma^{-1}\left(\frac{m_Q\omega m_Q|f|}{\lambda}\right) + C\gamma.\end{aligned}$$

Take $\gamma = \left(\frac{(m_Q\omega)m_Q|f|}{\lambda}\right)^{\frac{1}{p+1}}$. This gives (3.9) when $\lambda^{-1}(m_Q\omega)m_Q|f| \leq 1$, and the other case is trivial.

Let us use Lemma 3.7 to obtain good- λ and sharp function inequalities. We start with the analogue of Cotlar's inequality.

LEMMA 3.10. *Let T be as in Lemma 3.7, and set $T_*f(x) = \sup_{t>0} |\int_{|x-y|>t} K(x,y)f(y)dy|$. Then for $0 < \delta \leq 1$,*

$$T_*f \leq C[((|Tf|^\delta)^*)^{\frac{1}{\delta}} + \|T\|\lambda_\omega^{\beta,\epsilon}(f)]. \tag{3.11}$$

Here $\|T\|$ denotes the sum of the constants in the standard estimates (1.1) and (1.2) and in the estimate (3.8). As before, $\epsilon > 0$ comes from (1.2), and $\beta \in (0,1)$ is at our disposal.

The proof of this is similar to the classical case (see, e.g., p. 56 of [J]), using (3.3) and Lemma 3.7.

From these two lemmas we see that T_* satisfies (3.8) if T does, by taking $\delta < p$ in (3.11).

THEOREM 3.12. . *Assume that T is an ω -CZO (or just satisfies the hypotheses of Lemma 3.7). Then we have the following good- λ inequality: There is an $\eta > 0$ so that for $\gamma > 0$ small enough,*

$$|\{x : |T_*f| > 10\lambda, \lambda_\omega^{\beta,\epsilon}(f) \leq \gamma\lambda\}| \leq C\gamma^\eta|\{x : |T_*f| > \lambda\}|. \tag{3.13}$$

As always, $\varepsilon > 0$ comes from (1.2), while $\beta \in (0,1)$ is at our disposal.

The proof of (3.13) can be obtained in the standard way using (3.3) and the analogue of (3.9) with T replaced by T_*.

THEOREM 3.14. *Same assumptions as above. Given $q > 0$, define*

$$f_q^*(x) = \sup_{Q \ni x} \left(\frac{1}{|Q|} \int_Q |f|^q \right)^{\frac{1}{q}}, f_q^\sharp(x) = \sup_{Q \ni x} \inf_{c \in \mathbf{R}} \left(\frac{1}{|Q|} \int_Q |f - c|^q \right)^{\frac{1}{q}}.$$

Then: (a) there is a good- λ inequality in which the dyadic version of f_q^* is controlled by $f_q^\sharp$;

(b) $(Tf)_q^\sharp \leq C\lambda_\omega^{\beta,\varepsilon}(f)$ if q is small enough.

(If T satisfies (3.8), $q < \frac{p}{p+1}$ will work.)

The first part isOB proved just as in the case $q = 1$. (See [J], p. 41). The second part follows from the usual argument, using (3.3) and (3.9).

COROLLARY 3.15. *Same assumptions as above. Then T maps $L^p(\omega^p)$ into L^p and sends L^p into $L^p(\omega^{-p})$ boundedly for $1 < p < \infty$. Also, T maps L^∞ into BMO_ω.*

The L^p estimates follow from either of Theorems 3.12 or 3.14, using the corresponding L^p estimates for $\lambda_\omega^{\beta,\varepsilon}$. (For L^p into $L^p(\omega^{-p})$ we use the fact that $\omega^{-p} \in A_\infty$ for all $p < \infty$, since $\omega \in A_1$). The L^∞ into BMO_ω result can be obtained from the L^p estimates in the usual way, or one can show that $f_q^\sharp \leq C\omega$ implies $f_1^\sharp \leq C\omega$ for $\omega \in A_1$ (using a localized version of (a) in Theorem 3.14).

THEOREM 3.16. . *Same assumptions as above. Then T maps $L^1(\omega)$ into weak L^1*

This result is frustrating because $\lambda_\omega^{\beta,\varepsilon}(f)$ won't satisfy this weak-type result, suggesting that there should be a better choice of controlling maximal function. Note that (3.6) does satisfy such an estimate.

To prove this we use a Calderón-Zygmund decomposition. The tricky point is which $C-Z$ decomposition. By assumption, $f\omega \in L^1$. Given $\lambda > 0$, set $\Omega_\lambda = \{(\omega f)^* > \lambda\}$, which we can write as $\cup Q_i$, the Whitney decomposition of Ω_λ. Set $f = g + \Sigma b_i$, where $b_i = \chi_{Q_i}(f - m_{Q_i} f)$.

Choose $p > 1$ so that $m_Q \omega^p \le C(m_Q \omega)^p$. We claim that $\int |g|^p \omega^p \le C\lambda^{p-1} \int |f|\omega$. Indeed, on $\mathbf{R}^n \backslash \Omega_\lambda$ we have that $\omega|g| = \omega|f| \le \lambda$, so that

$$\int_{\mathbf{R}^n \backslash \Omega_\lambda} |g|^p \omega^p \le \lambda^{p-1} \int_{\mathbf{R} \backslash \Omega_\lambda} |f|\omega.$$

Also,

$$\begin{aligned}
\int_{\Omega_\lambda} |g|^p \omega^p &= \Sigma_i \int_{Q_i} (m_{Q_i}|f|)^p \omega^p \\
&= \Sigma_i (m_{Q_i}|f|)^p |Q_i| m_{Q_i}(\omega^p) \\
&\le C\Sigma_i (m_{Q_i}|f|)^p |Q_i| (m_{Q_i}\omega)^p \\
&\le C\Sigma_i (m_{Q_i}|f|\omega)^p |Q_i| \quad \text{since} \quad \omega \in A_1 \\
&\le C\lambda^{p-1} \Sigma_i m_{Q_i}(|f|\omega)|Q_i| = C\lambda^{p-1} \int_{\Omega_\lambda} |f|\omega
\end{aligned}$$

For the last inequality we used $m_{Q_i}(|f|\omega) \le C\lambda$, which comes from the fact that dist $(Q_i, \mathbf{R}^n \backslash \Omega_\lambda) \le C|Q_i|^{\frac{1}{n}}$.

Thus

$$|\{x : |Tg| > \frac{\lambda}{2}\}| \le C\lambda^{-p} \int |Tg|^p \le C\lambda^{-p} \int |g|^p \omega^p \le C\lambda^{-1} \int |f|\omega.$$

Since $m_{Q_i}(|f|\omega) \le \lambda$, we have that $m_{Q_i}|b_i| \le 2m_{Q_i}|f| \le 2m_{Q_i}(|f|\omega)(m_{Q_i}\omega)^{-1} \le 2\lambda\omega(Q_i)^{-1}$. Hence from (1.2) we get (setting $y_i =$ center Q_i),

$$\begin{aligned}
\int_{\mathbf{R}^n \backslash 2Q_i} |Tb_i| &\le C \int_{\mathbf{R}^n \backslash 2Q_i} \int_{Q_i} m_{Q_i}\omega \frac{|Q_i|^{\frac{\varepsilon}{n}}}{|x - y_i|^{n+\varepsilon}} |b_i(t)| dt dx \\
&\le C m_{Q_i}\omega |Q_i| m_{Q_i}|b_i| \\
&\le C\lambda|Q_i|.
\end{aligned}$$

Put $E_\lambda = U(2Q_i)$. Then

$$\int_{\mathbf{R}^n \backslash E_\lambda} |T(\Sigma b_i)| \le \Sigma \int_{\mathbf{R}^n \backslash 2Q_i} |T(b_i)| \le C\lambda|\Omega_\lambda| \le C \int |f|\omega.$$

Altogether,

$$|\{x : |Tf(x)| > \lambda\}| \leq |\{|Tg| > \frac{\lambda}{2}\}| + |T(\Sigma b_i)| > \{\frac{\lambda}{2}\}|$$
$$\leq C\lambda^{-1} \int |f|\omega + |E_\lambda| + |\{x \notin E_\lambda |T(\sum_i b_i)| > \frac{\lambda}{2}\}|$$
$$\leq C\lambda^{-1} \int |f|\omega + 2|\Omega_\lambda| + C\lambda^{-1} \int_{\mathbf{R}^n \setminus E_\lambda} |T(\sigma b_i)|$$
$$\leq C\lambda^{-1} \int |f|\omega.$$

This proves Theorem 3.16.

References

[CM1]. R.R. Coifman and Y. Meyer, *Au-delà des opérateurs pseudodifferéntiels*, Asterisque **57** (1978).

[CM2]. R.R. Coifman and Y Meyer, *Une généralisation du théoréme de Calderón sur l'intégrale de Cauchy*; In Fourier Analysis, proceedings of the seminar of El Escorial, June 1979, Asoc. Mat. Española, Madrid (1980).

[CRW]. R. Coifman, R. Rrochberg, and G. Weiss, *Factorization theorems for Hardy spaces in several variables*, Ann. Math **103** (1976), 611-635.

[GC]. J. Garcia-Cuerva, *Weighted Hardy spaces,*; In *Harmonic Analysis in Euclidean Spaces* Proc. Symp. Pure Math (AMS) 35 **I** (1979).

[J]. J.L. Journé, *Calderón-Zygmund operators, Pseudodifferential operators, and the Cauchy integral of Calderón*, Springer Lecture Notes **994** (1983).

[Ja]. S. Janson, *Mean oscillation and commutators of singular integral operators*, Arkiv för Math **16** (1978), 263-270.

[L]. M. Leckband, *On the local boundedness of some singular integral operators*, preprint (1986).

On the $\bar{\partial}_b$ equation and Szegö projection on CR manifolds

Mike Christ[1]

University of California, Los Angeles
Department of Mathematics
Los Angeles, California 90024

0. Prelude.

This article is an announcement of results concerning the regularity properties of solutions of a certain differential equation which arises naturally in complex analysis in several variables, obtained in a series of works [**C2**],[**C3**],[**C5**]. The relevant definitions are given, several theorems are stated, some ingredients of the proofs are discussed and finally the overall structure of the proofs is briefly sketched. Details will appear in the papers cited.

This investigation began as joint work with J.J. Kohn, and relies in an essential way on an idea of his. The analysis also depends on earlier work of Rothschild and Stein [**RS**] and Nagel, Stein and Wainger [**NSW**]. Closely related results have been obtained independently by Kohn and C. Fefferman [**FK**]. Applications to complex analysis may be found in [**FK**] and [**K3**]. We are of course indebted to Kohn, and also to Fefferman for one encouraging conversation and A. Nagel for several.

1. Definitions.

Let M be a compact, C^∞ manifold without boundary, of dimension 3 over $\mathbb{R}$. Let there be given a C^∞ complex sub–bundle $T^{1,0}$ of its complexified tangent bundle, with fibers $T^{1,0}_x$ of dimension 1 over $\mathbb{C}$, and suppose that $T^{1,0}_x \cap \bar{T}^{1,0}_x = \{0\}$ for all $x \in M$. Then M is said to be a CR manifold. (The notion of CR manifold exists in higher dimensions, but our results are confined to the case of real dimension three.) Let $B^{0,1}$ be the bundle dual to $\bar{T}^{1,0}_x$; its fibers also have dimension 1 over $\mathbb{C}$. Associated to $B^{0,1}$ is a canonical first–order differential operator, $\bar{\partial}_b$, which maps smooth functions to sections of $B^{0,1}$. $\bar{\partial}_b u(x)$ is the restriction of $du(x)$ to a linear functional on $\bar{T}^{1,0}_x$, or in other words if $\bar{Z} \in \bar{T}^{1,0}_x$ then $\bar{\partial}_b u(\bar{Z}) = \bar{Z}u(x)$. In local coordinates in a coordinate patch $U \subset M$, $B^{0,1}$ may be identified with $U \times \mathbb{C}$, and $\bar{\partial}_b$ becomes a complex vector field $X + iY$ where X, Y are C^∞ real vector fields, linearly independent at every point.

The boundary M of any open, relatively compact smoothly bounded domain $\Omega \subset \mathbb{C}^2$ has a natural CR structure. $T^{1,0}_x$ is the (complex one–dimensional) space of all holomorphic vectors $a\frac{\partial}{\partial z_1} + b\frac{\partial}{\partial z_2}$ tangent to M at x, while $B^{0,1}$ is the set of all restrictions to $\bar{T}^{1,0}_x$ of

[1]Research supported in part by NSF grants and the Mathematical Sciences Research Institute.

$(0,1)$ forms on $\mathbf{C}^2$. In this case $\bar\partial_b$ is directly tied to complex analysis on Ω, and it may be given an alternative definition as the composition

$$u \longrightarrow \tilde{u} \longrightarrow \bar\partial\tilde{u} \longrightarrow \bar\partial\tilde{u}|_M$$

where $u \in C^1(M)$ is any function given, $\tilde{u}$ is an arbitrarily chosen extension of u to a function in $C^1(\bar\Omega)$, $\bar\partial$ is the ordinary Cauchy–Riemann operator mapping functions to $(0,1)$ forms in $\mathbf{C}^2$, and the restriction of $\bar\partial\tilde{u}$ to the boundary is taken in the sense of differential forms, as a section of the dual bundle of $\bar{T}^{1,0}$. The end result is independent of the choice of the extension. However not all CR manifolds arise as boundaries.

Let $\mathcal{H}_b$ denote the kernel of $\bar\partial_b$ in $L^2(M)$, a closed subspace of $L^2(M)$. When $M = \partial\Omega \subset \mathbf{C}^2$, $\mathcal{H}_b$ contains the restriction to M of every holomorphic function in $C^1(\bar\Omega)$, as is clear from the alternative definition of $\bar\partial_b$ in terms of extensions, so $\mathcal{H}_b$ has infinite dimension. In fact $\mathcal{H}_b$ equals H^2, the set of all L^2 functions on M whose harmonic extensions to Ω are holomorphic.

The Szegõ projection S is the orthogonal projection from $L^2(M,\mu)$ to $\mathcal{H}_b$, with respect to any fixed measure μ with $d\mu = a\,dx$ in local coordinates, $a \in C^\infty$ strictly positive. It is analogous to the Cauchy projection of L^2 onto H^2 on the circle in the complex plane $\mathbf{C}^1$.

The results to be discussed in this article concern regularity properties of solutions of the differential equation $\bar\partial_b u = f$ and mapping properties of the Szegõ projection.

2. Hypotheses.

Suppose $u, f \in L^2(M)$, more precisely that u is an L^2 function and f an L^2 section of $B^{0,1}$, and that $\bar\partial_b u = f$. One might hope that u must then be somewhat smoother than f. But consider the case $f = 0$. If $\mathcal{H}_b$ has infinite dimension, as is typical, then by elementary functional analysis it cannot be contained in any space such as a Sobolev space L^2_ϵ, $\epsilon > 0$, which embeds compactly in L^2. Thus even when $f \in C^\infty$ one cannot conclude that $u \in L^2_\epsilon$, and hence there can be no regularity theory for arbitrary solutions of $\bar\partial_b u = f$.

Besides the infinite–dimensional kernel we may cite two further indications that the $\bar\partial_b$ equation is poorly behaved. First, it is in general not even locally solvable (and in fact is never locally solvable on the subclass of CR manifolds which we shall study); given a typical $x_0 \in M$ and typical $f \in C^\infty$ in a neighborhood of x_0, no neighborhood of x_0 may be found in which a solution u exists. H. Lewy [L] established this for the most fundamental example of a CR manifold, the Heisenberg group. Second, $\bar\partial_b$ is not elliptic. In local coordinates $\bar\partial_b = X + iY$, but we are working on a 3–dimensional manifold so that one direction remains unaccounted for. Certainly to control $\bar\partial_b u$ is no better than to control Xu and Yu separately.

Following Kohn, we shall impose three hypotheses on M, one of a global analytic and two of a local geometric character. First we assume that $\bar\partial_b$ has closed range on L^2. That is, if f belongs to the closure in L^2 of $L^2 \cap \bar\partial_b(L^2)$, then there exists $u \in L^2$ satisfying $\bar\partial_b u = f$, and u may be chosen so that $\|u\|_2 \leq C\|f\|_2$ where C depends only on M. Kohn [**K1**] has proved that this holds automatically whenever M is the boundary of a pseudoconvex domain $\Omega \subset \mathbf{C}^2$. In general the issue of closed range is closely related to the question of global CR embeddability of M in some $\mathbf{C}^n$.

The second hypothesis is that M is pseudoconvex. To define this notion consider a coordinate patch $U \subset \mathbf{R}^3$ on which $\bar{\partial}_b = X + iY$ and on which X, Y and $T = \frac{\partial}{\partial x_3}$ are everywhere linearly independent. Then it is possible to express the Lie bracket $[X, Y]$ as

$$[X,Y](x) = \lambda(x)T + a(x)X + b(x)Y \tag{2.1}$$

where $\lambda, a, b \in C^\infty(U)$. We say that M is pseudoconvex if it may be covered by such coordinate patches, in each of which λ has constant sign. We shall always choose the local orientation so that

$$\lambda(x) \geq 0 \quad \text{for all } x.$$

Except for this sign convention, the notion of pseudoconvexity is independent of the choice of local coordinates. We say that U is pseudoconvex if the above holds in U, rather than on all coordinate patches. In the case where $M = \partial\Omega \subset \mathbf{C}^2$, pseudoconvexity of M is equivalent to pseudoconvexity of Ω in the usual sense of complex analysis—in other words, Ω is a domain of holomorphy. Thus this hypothesis is natural from the perspective of complex analysis. M is said to be strongly pseudoconvex if, in local coordinates, λ never vanishes, and weakly pseudoconvex if it is pseudoconvex but λ does vanish somewhere. The strongly pseudoconvex case is already comparatively well understood, and it is the weakly pseudoconvex case which is to be studied here.

The third hypothesis is that M should be of finite type. This means that in a coordinate patch as above, the vector fields

$$X,\ Y,\ [X,Y],\ [X,[X,Y]],\ [Y,[X,Y]],\ \ldots,$$

that is all vector fields generated by taking iterated commutators of any order of X and Y, should span the tangent space at each point. Since X and Y are linearly independent, at a point x this means that some one of the commutators must be linearly independent of X and Y. Let $m(x)$ be the minimum of the lengths of such linearly independent commutators at x, where $[X,Y]$ is defined to have length 2, $[X,[X,Y]]$ to have length 3, and so on. $m(x)$ is independent of the choice of coordinates. The type of M is defined to be

$$m = \max_{x \in M} m(x).$$

Similarly we may speak of the type of an open set U. It is well known that under the finite type hypothesis, $\Box = -X^2 - Y^2$ is subelliptic.

Finally in order to have a chance of overcoming the obstacle of the infinite–dimensional kernel, we consider, following Kohn, the equation

(*)
$$\bar{\partial}_b u = f$$
$$u \perp \mathcal{H}_b.$$

By the closed range hypothesis, for each $f \in \text{Range}(\bar{\partial}_b) = L^2 \cap \bar{\partial}_b(L^2)$ there exists a unique $u \in L^2$ satisfying (*), and $\|u\|_2 \leq C\|f\|_2$.

Denote by L^p_s the Sobolev space of all functions having s derivatives in L^p. Kohn [**K2**] has shown that under the assumptions of pseudoconvexity, closed range and finite type, there exists $\epsilon > 0$ (in fact $\epsilon = m^{-1}$) such that for all $s \geq 0$, for any $f \in L^2_s(M) \cap \text{Range}(\bar{\partial}_b)$ the unique solution u of (*) belongs to $L^2_{s+\epsilon}(M)$. Moreover if f belongs to L^2_s on some open subset V of M and is globally in L^2, then $u \in L^2_{s+\epsilon}$ on any compact subset of V. The finite type hypothesis is necessary for any result of this type, with a gain of some fraction of a derivative, to be valid. More quantitatively it is necessary that $\epsilon \leq m^{-1}$.

3. Some regularity results.

The equation (*) has global aspects, namely the condition $u \perp \mathcal{H}_b$ and the assumptions that f belongs to $L^2(M)$ and that $\bar{\partial}_b u = f$ on all of M. However its analysis may be based on a purely local situation. Let V be an open ball in $\mathbb{R}^3$, and X, Y be real, C^∞ vector fields such that X, Y and $T = \frac{\partial}{\partial x_3}$ are linearly independent at each point of V. Suppose

$$[X,Y] = \lambda(x)T + O(X,Y)$$

at each $x \in V$, where $O(X,Y)$ denotes a linear combination of $X(x)$ and $Y(x)$, and $\lambda(x) \geq 0$ for all $x \in V$. Suppose moreover that the pair (X,Y) is of finite type m on V, as defined above. Let $\epsilon = m^{-1}$. Let $\bar{Z}$ be a first order differential operator of the form

$$\bar{Z}(u) = (X + iY)u(x) + a(x)u(x)$$

where $a \in C^\infty(V)$. Let

$$Zu(x) = (X - iY)u(x) + b(x)u(x)$$

where $b \in C^\infty(V)$ need not be related in any way to a. Suppose that on V,

(3.1)
$$\bar{Z}u = f$$
$$Zv = u.$$

THEOREM A. *For each compact subregion $K \subset V$ and each $p \in (1,\infty)$, $q > 1$ and $s \geq 0$ there exists $C < \infty$ such that for all $u, v \in L^q(V)$ and $f \in L^p_s(V)$ satisfying (3.1),*

$$\|u\|_{L^p_s(K)} + \|Xu\|_{L^p_s(K)} + \|Yu\|_{L^p_s(K)} \leq C\big(\|f\|_{L^p_s(V)} + \|u\|_{L^q(V)} + \|v\|_{L^q(V)}\big).$$

COROLLARY B. *Under the same hypothesis*

$$\|u\|_{L^p_{s+\epsilon}(K)} \leq C\big(\|f\|_{L^p_s(V)} + \|u\|_{L^q(V)} + \|v\|_{L^q(V)}\big).$$

Thus the optimal gain $\epsilon = m^{-1}$ is obtained. The corollary follows from the theorem by the work of Rothschild and Stein [RS].

Now the closed range hypothesis reduces the study of (*) to the situation of Theorem A. For $\bar{\partial}_b^*$, the adjoint, has closed range since $\bar{\partial}_b$ does, and $\mathcal{H}_b^\perp = \text{Range}(\bar{\partial}_b^*) = L^2 \cap \bar{\partial}_b^*(L^2)$ (we do not distinguish in the notation between L^2 functions and L^2 sections of $B^{0,1}$). Thus any $u \perp \mathcal{H}_b$ may be written as $u = \bar{\partial}_b^* v$, with $\|v\|_2 \leq C\|u\|_2$, so that we are really studying the equation

$$(3.2) \qquad\qquad \bar{\partial}_b(\bar{\partial}_b^* v) = f$$

given that $\|v\|_2 + \|\bar{\partial}_b^* v\|_2 \leq C\|f\|_2$, where it is $\bar{\partial}_b^* v$ rather than v itself whose regularity is to be investigated. In local coordinates in which $\bar{\partial}_b = X + iY$, $\bar{\partial}_b^*$ becomes $X - iY + b(x)$, where $b \in C^\infty$. One advantage in dealing with (3.2) is that it is essentially local in character. This point of view is that of Kohn [K2]. Thus we obtain

THEOREM C. *Let M be a CR manifold of dimension 3 on which $\bar{\partial}_b$ has closed range. Let $p \in (1, \infty)$. Let $f \in \text{Range}(\bar{\partial}_b)$. Suppose that U is a pseudoconvex open subset of finite type m, and that $f \in L^p_s(U)$. Then the unique solution u of (*) belongs to $L^p_{s+\epsilon}$ on every compact subregion of U, where $\epsilon = m^{-1}$.*

It is not necessary to assume that M is globally pseudoconvex or of finite type. Fefferman and Kohn [FK] have obtained the analogous result for the scale of Hölder classes. (Just by taking p very large in Theorem C we may conclude that there is a gain of any order strictly less than m^{-1} in terms of Hölder regularity.)

Concerning the Szegő projection we have

THEOREM D. *The Szegő projection extends to an operator bounded on L^p for all $p \in (1, \infty)$.*

This is in accord with the analogy between the Szegő projection and the Cauchy projection in one complex variable. Actually the analogy goes further—the Szegő projection is a Calderón–Zygmund singular operator, with respect to a natural geometric structure on M. See Theorem F below. The connection between S and (*) is quite simple: If $g \in C^1$, then $(I - S)g$ is the solution u of (*), with $f = \bar{\partial}_b g$. Thus Theorem D is just another regularity result for (*).

4. Pointwise bounds.

Nagel, Stein and Wainger [NSW] have shown how the CR structure induces naturally the structure of a space of homogeneous type on M, in the sense of Coifman and Weiss

[CW], under the finite type hypothesis. Fix a small coordinate patch $U \subset M$ and write $\bar\partial_b = X + iY$ in local coordinates as above. Let K be any fixed compact subset of U. Assume that $\bar\partial_b$ is of type at most m on U. For each commutator

$$V = [(X \text{ or } Y),\ [(X \text{ or } Y),\ [\ldots,(X \text{ or } Y)]\ldots]$$

of any length $2 \leq j \leq m$, define

$$\Lambda_V(x,r) = r^{2+j}\left|\text{determinant}\{X(x),Y(x),V(x)\}\right|$$

for $(x,r) \in K \times \mathrm{R}^+$. Set

$$\Lambda(x,r) = \sum \Lambda_V(x,r),$$

summed over all such V. $\Lambda(x,r)$ is a polynomial of degree $m+2$ in r, for each x. To each pair (x,r) associate some commutator $W = W(x,r)$ satisfying

$$\Lambda_W(x,r) \geq \frac{1}{2}\max_V \Lambda_V(x,r),$$

where only commutators of lengths less than or equal to m are considered. Let $\iota(x,r)$ be the length of $W(x,r)$. Let $\hat B = \{\xi \in \mathrm{R}^3 : |\xi| < 1\}$ and define $\phi = \phi_{x,r} : \hat B \mapsto U$ by

$$\phi(\xi) = \exp(c_0(r\xi_1 X + r\xi_2 Y + r^{\iota}\xi_3 W))x$$

where $\iota = \iota(x,r)$ and $W = W(x,r)$, and where c_0 is some sufficiently small constant. Define

$$B(x,r) = \phi(\hat B)$$

and

$$\rho(x,y) = \inf\{r : y \in B(x,r)\}.$$

The following fundamental result of [NSW] is valid provided U, c_0 and $\delta > 0$ are chosen sufficiently small.

THEOREM. *For all $x,y,z \in K$ and $r \in (0,\delta]$:*

(1) $\phi : \hat B \mapsto U$ *is a diffeomorphism onto an open subset.*

(2) $0 < \rho(x,y) < \infty$ *for all* $x \neq y$.

(3) $\rho(x,y) = \rho(y,x)$.

(4) $\rho(x,y) \leq C(\rho(x,z) + \rho(z,y))$.

(5) $|B(x,r)| \sim \Lambda(x,r)$.

The metric ρ is independent of all choices made in the construction, up to a factor bounded above and below. This makes K into a space of homogeneous type, and patching together finitely many coordinate charts does the same for M.

The vector fields X, Y pull back via ϕ to vector fields $\hat X, \hat Y$ on $\hat B$ which are also of finite type, and whose coefficients are C^∞ functions on $\hat B$; both properties hold uniformly in x, r. The equation $(X+iY)u = f$ becomes $(\hat X + i\hat Y)\hat u = r\hat f$ where $\hat u = u \circ \phi$ and $\hat f = f \circ \phi$. The presence of the factor of r in the rescaled equation suggests the following fractional integration inequality.

PROPOSITION E. *Let $f \in Range(\bar{\partial}_b)$. Then the unique solution u of (*) satisfies*

$$\|u\|_{L^2(B(x,r))} \leq Cr\|f\|_{L^2(M)}$$

for all x, r. C depends only on M.

We assume implicitly throughout this section that M is pseudoconvex of finite type and that $\bar{\partial}_b$ has closed range, but all the results stated here have local versions as in Theorem C.

The next two theorems are obtained by combining first, Proposition E, second, Theorem A with uniform control of the bounds involved, third, the closed range property, and finally the rescaling maps $\phi_{x,r}$ with $r \sim \rho(x,y)$. D denotes any differential operator in both the variables x, y of the form $D = (X \text{ or } Y) \circ (X \text{ or } Y) \circ \ldots$ with n factors, with each (X or Y) acting in either the x or the y variable.

THEOREM F. *The distribution-kernel K for the Szegő projection is C^∞ off of the diagonal and satisfies*

$$|K(x,y)| \leq C\Lambda(x,\rho)^{-1}$$

and more generally

$$|DK(x,y)| \leq C_n\rho^{-n}\Lambda(x,\rho)^{-1}$$

where $\rho = \rho(x,y)$.

Consider next the relative fundamental solution for (*). This is the operator which first projects any L^2 section f of $B^{0,1}$ orthogonally onto Range($\bar{\partial}_b$), then maps to the unique solution of (*).

THEOREM G. *The distribution-kernel $L(x,y)$ for the relative fundamental solution of (*) is C^∞ away from the diagonal and satisfies*

$$|L(x,y)| \leq C\rho\Lambda(x,\rho)^{-1}$$

and more generally

$$|DL(x,y)| \leq C_n\rho^{1-n}\Lambda(x,\rho)^{-1}$$

where $\rho = \rho(x,y)$. Moreover the operator is realized on L^2 functions by integration against L (almost everywhere).

An easy corollary is

THEOREM H. *Let $f \in Range(\bar{\partial}_b)$ belong to L^∞ on some open set U. Then the unique solution u of (*) is Hölder continuous of order m^{-1} on each compact subset of U. Moreover*

$$|u(x) - u(y)| \leq C\rho(x,y)\log(2 + \rho(x,y)^{-1})$$

for all x, y contained in any compact subset.

Moreover Xu and Yu belong to a certain space BMO, on compact subsets of U, defined with respect to the balls $B(x,r)$. In the strongly pseudoconvex case stronger versions of Theorems F and G are known and are due to Fefferman [F] (see also [BS]), and to Folland and Stein [FS] respectively. In the weakly pseudoconvex case certain special (model) manifolds M have been treated in [N], [M] and [NRSW]. More recently Theorem F has also been obtained by Nagel, Stein, Rosay and Wainger in the case of a boundary in $\mathbf{C}^2$ [NS].

Theorem D may be deduced from Theorem F and standard Calderón–Zygmund theory, using further properties of the balls and rescaling maps.

The potential relevance of rescaling arguments in the context of Theorem F was pointed out to this author long ago by A. Nagel.

5. Inverses of singular integral operators.

This section and the next are devoted to discussions of two relatively nonstandard ingredients used in the proof of Theorem A. Let $\mathbf{g}$ be a finite-dimensional Lie algebra of the form

$$\mathbf{g} = \bigoplus_j \mathbf{g}_j$$

as a vector space, where

$$[\mathbf{g}_i, \mathbf{g}_j] \subset \mathbf{g}_{i+j}$$

for all i, j, and where $\mathbf{g}_1$ generates $\mathbf{g}$ as a Lie algebra. Let G be the associated connected, simply connected nilpotent Lie group and identify elements of $\mathbf{g}$ with left-invariant vector fields on G via the exponential map. The homogeneous dimension of G is defined to be $d = \sum j \cdot \text{dimension}(\mathbf{g}_j)$. Let 0 denote the group identity element.

Fix a basis $\{X_j\}$ for $\mathbf{g}_1$ (as a vector space). For each $r > 0$ there is a unique Lie algebra automorphism δ_r of $\mathbf{g}$ satisfying $\delta_r X_j = r X_j$ for each j. The identification of $\mathbf{g}$ with G via the exponential map defines corresponding group automorphisms of G, which we also denote by δ_r. They play the role of dilations in Euclidean harmonic analysis.

Form the sub-Laplacian

$$\Box = -\sum X_j^2,$$

a left-invariant subelliptic differential operator on G. For $s \geq 0$ and $q \in (1, \infty)$ define the anisotropic Sobolev space

$$\mathcal{L}_s^q = \{f \in L^q(G) : \Box^{s/2} f \in L^q\}.$$

See [Fo] for a discussion of the powers of $\Box$ and [RS] for the basic properties of these spaces (at least when $s \in \mathbf{Z}^+$).

We say that a distribution K is homogeneous of degree $-d$ if, setting $f_r(x) = f(\delta_r x)$ for all test functions f,

$$\langle K,\ f_r\rangle = \langle K,\ f\rangle$$

for all $f \in \mathcal{S}$ and $r > 0$. Every such distribution is a linear combination of the Dirac measure at 0 plus a "principal-value" distribution; see [**C2**] for the precise statement. We say that $K \in \mathcal{L}^q_s$ away from 0 if for all $\eta \in C_0^\infty(G)$ vanishing identically in a neighborhood of 0, $\eta K \in \mathcal{L}^q_s$.

Let $\mathcal{A}_{q,s}$ denote the class of all convolution operators $Tf = f * K$, wtih K homogeneous of degree $-d$ and $K \in \mathcal{L}^q_s$ away from 0. In the simplest case where $G = \mathbf{R}^n$ with the usual dilation structure, this means $K = c\delta_{x=0} + K'$ where K' is homogeneous of degree $-n$, and has mean value 0 and belongs to the ordinary Sobolev space L^q_s on the unit sphere.

PROPOSITION I. *For each $q > 1$ and $s \geq 0$, every element of $\mathcal{A}_{q,s}$ extends to an operator bounded on $L^p(G)$ for every $p \in (1, \infty)$.*

The case $s > 0$ relies on the Cotlar–Stein almost-orthogonality lemma and standard Calderón–Zygmund techniques, as in [**KS**]. The case $s = 0$ [**C4**] is more subtle and relies on the method of [**C1**].

$\mathcal{A}_{q,s}$ turns out to be an algebra, and may be given a norm under which it becomes a Banach algebra (for it suffices to take a single η in the definition, so long as η vanishes nowhere on the unit sphere). The following is our main result concerning it.

THEOREM J. [**C2**],[**C4**] *Let $q \in (1, \infty)$ and $s \geq 0$. Suppose that $T \in \mathcal{A}_{q,s}$ is invertible as an operator on $L^2(G)$. Then $T^{-1} \in \mathcal{A}_{q,s}$ also.*

The question in the endpoint case was raised by A. Carbery. Other results of this general type, asserting that certain Banach algebras of bounded operators on some L^2 are closed under inversion, are Wiener's Tauberian theorem and R. Beals' theorem on the class $S^0_{\frac{1}{2},\frac{1}{2}}$ of pseudodifferential operators. The C^∞ case of Theorem J was obtained some time ago by [**CG**], but the case when q is close to 1 and s to 0 required a new method. In the case $s = 0$, the theorem is valid more generally, on any connected, simply connected nilpotent Lie group equipped with a suitably nondegenerate one-parameter group of automorphic dilations. The Lie algebra need not possess the stratified structure assumed above.

It is straightforward to verify that any element of the algebra $\mathcal{A}_{q,s}$ is bounded on $\mathcal{L}^q_s$. Hence a consequence of the theorem is that T is invertible on $\mathcal{L}^q_s$. On the other hand it is a comparatively easy step to show that if $T \in \mathcal{A}_{q,s}$ is invertible on $\mathcal{A}_{q,s}$, then T^{-1} belongs to $\mathcal{A}_{q,s}$. The proof in [**C2**] proceeded in this way, first establishing the invertibility on $\mathcal{A}_{q,s}$. This is a curious reversal of the usual procedure in Calderón–Zygmund theory; instead of using properties of the convolution kernel of T^{-1}, a bounded operator on L^2, to verify that it is bounded on other function spaces, we prove it is bounded on other spaces in order to establish regularity of its convolution kernel.

In practice it can actually happen that invertibility on L^2 is easy to verify. Our application of Theorem J is to

$$T = I + \Box^{-1/2}|[X,Y]|\Box^{-1/2}$$

where I is the identity operator, $\mathfrak{g}$ is the free nilpotent Lie algebra of step m on two generators X and Y, and the absolute value of $[X,Y]$ is as defined below. It is pretty clear that T is a self-adjoint, nonnegative perturbation of the identity, and it can be shown to belong to some $\mathcal{A}_{q,s}$, with essentially a precise value of s. Therefore it is bounded on L^2 by Proposition I, and it is invertible since it is $\geq I$.

In our eventual application of Theorem J it is essential to know that no derivatives at all are lost in the passage from the convolution kernel for T to that for T^{-1}.

6. The absolute value of a vector field.

We discuss briefly two definitions which we found to be quite useful in the proof of Theorem A. First recall that on $\mathbf{R}^1$, one may define $|\frac{d}{dx}|$ by Fourier transformation:

$$(\left|\frac{d}{dx}\right| f)\hat{}(\xi) = 2\pi|\xi|\hat{f}(\xi).$$

There exists a distribution $\mathcal{D}$, homogeneous of degree -2 on $\mathbf{R}$ and C^∞ away from the origin, so that $\hat{\mathcal{D}}(\xi) = 2\pi|\xi|$. Thus

$$\left|\frac{d}{dx}\right| f(0) = \langle f, \mathcal{D}\rangle$$

for all $f \in \mathcal{S}$.

In $\mathbf{R}^n$ we may define the absolute value of a coordinate vector field in the same manner. More generally for a real, C^∞ vector field which never vanishes we might first adopt coordinates in which it becomes (locally) a coordinate vector field, then invoke the Fourier transform. However our proof of Theorem A depends on manipulating the absolute value of $[X,Y]$, which will vanish at some points in the weakly pseudoconvex case of interest.

Therefore consider an arbitrary real, C^∞ vector field V defined in some open set in $\mathbf{R}^n$. For $f \in C^\infty$ define

$$|V|f(x) = \langle g, \mathcal{D}\rangle$$

where $g \in C_0^\infty(\mathbf{R}^1)$ is defined by

$$g(t) = \varphi(t) f(\exp(tV))(x),$$

exp denotes the exponential map, and $\varphi \in C_0^\infty(\mathbf{R})$ is an auxiliary function which is identically one in some neighborhood of 0. Choosing a different φ results in a perturbation by an operator bounded on any L^p and on any (Euclidean) Sobolev space, so the definintion is essentially independent of φ. Its principal advantages are that it is at least well-defined for any V, provided φ is chosen to have small enough support that the exponential map is defined, that it is defined directly in terms of the geometric behavior of V, and that it is diffeomorphism-invariant.

More generally suppose that $U \subset \mathbf{R}^n$ is open and that $\gamma : U \times \mathbf{R} \mapsto \mathbf{R}^n$ is C^∞ and satisfies $\gamma(x,0) \equiv x$. An example is $\gamma(x,t) = \exp(tV)x$ where V is any vector field, but not every C^∞ family of curves γ arises in this way. Define

$$|D_\gamma| f(x) = \langle g, \mathcal{D} \rangle$$

where now

$$g(t) = \varphi(t)\, f(\gamma(x,t)),$$

for x in any fixed compact subset of U, where φ has small support.

The cutoff function φ is required in part because $\exp(tV)(x)$ will in general be defined for small t. In certain cases, however, the exponential map is defined for all t and it is more natural to omit φ in the definition of $|V|$. In particular this occurs when V is a coordinate vector field in $\mathbf{R}^n$, or more generally a left-invariant vector field on a group G of the type described above.

7. Outline of the proof.

There are five principal steps in the proof of Theorem A. The first, due entirely to Kohn, is to use a microlocal analysis to reduce the problem to that of obtaining a parametrix $\mathcal{P}$ for the operator

$$A = \square + \lambda(x)|T|,$$

where

$$T = \frac{\partial}{\partial x_3} \quad \text{and} \quad \square = -(X^2 + Y^2),$$

with suitably strong mapping properties; in particular $X \circ \mathcal{P}$ and $Y \circ \mathcal{P}$ should be bounded on all L^p. The term $\lambda(x)\left|\frac{\partial}{\partial x_3}\right|$ denotes the composition of the absolute value of the coordinate vector field, followed by multiplication by the function λ.

To see how the reduction goes consider a coordinate patch as above. Let P^+ be a translation-invariant classical pseudodifferential operator of order zero whose symbol is supported on the cone $\{|\xi_3| > C|(\xi_1, \xi_2)|\} \cap \{\xi_3 > 0\}$. Differentiate the equation $\bar\partial_b u = f$ to obtain $\bar\partial_b^* \bar\partial_b u = \bar\partial_b^* f$. By (2.1),

$$\bar\partial_b^* \bar\partial_b = -\square + i[X,Y] + O(X,Y,I) = -\square + i\lambda(x)T + O(X,Y,I),$$

and the symbol of iT is $-2\pi\xi_3 = -2\pi|\xi_3|$ on the microlocal support of P^+. Thus $\bar\partial_b^* \bar\partial_b \circ P^+ = (-A + O(X,Y,I)) \circ P^+$. Moreover by the usual calculus of pseudodifferential operators, $\bar\partial_b^* \bar\partial_b P^+ u = \bar\partial_b^* P^+ \bar\partial_b u$ modulo a lower-order error term. Thus

$$-A(P^+ u) = \bar\partial_b^* P^+ f$$

modulo error terms, so that P^+u may be recovered once a parametrix for A is known.

On the region $\{|\xi_3| \le C|(\xi_1,\xi_2)|\}$, $\bar\partial_b^*\bar\partial_b$ is a second-order elliptic operator, so the standard elliptic theory applies. Finally the microlocal region where $|\xi_3| \ge C|(\xi_1,\xi_2)|$ may be treated by a variant of the first case; the same operator A arises. It is here that the representation $u = \bar\partial_b^* v$ is used. This argument is almost implicit in [**K2**].

The second step is to show that essentially $\lambda(x)|T|$ equals $|[X,Y]|$ modulo a lower-order error term. $\lambda(x)|T|$ is $|D_\gamma|$ where $\gamma(x,t) = (x_1, x_2, x_3 + \lambda(x)t)$. γ is not the exponential map associated to any vector field, but it may be related in a natural way to the vector fields X, Y and $[X,Y]$ using an elegant idea of Nagel, Stein and Wainger [**NSW**] (see the article by Professor Wainger elsewhere in these proceedings).

The next step is to prove Theorem J (in the easier case $s > 0$). The original proof [**C2**] was rather involved, but we shall give a much simpler argument in a forthcoming publication [**C6**].

Step four is to analyze the regularity of the convolution kernel for the invariant operator

$$\hat{\Box} + |[\hat X, \hat Y]|$$

where $\hat{\Box} = -\hat X^2 - \hat Y^2$ on G, the free nilpotent group of step m on 2 generators $\hat X$ and $\hat Y$. This is done by applying Theorem J to

$$I + \hat{\Box}^{-1/2}|[\hat X, \hat Y]|\hat{\Box}^{-1/2},$$

which turns out to belong to A_s^q for some $s, q > 1$. Therefore it is bounded on L^2. It is a self-adjoint and nonnegative perturbation of the identity, so is invertible on L^2.

Most of the labor goes into the fifth and last step. Let X, Y be the original vector fields on a coordinate patch in M. Rothschild and Stein [**RS**], building on earlier work of [**FS**], developed an ingenious method for studying the regularity properties of $\Box$. They first understand the regularity of the convolution kernel for the inverse of the model operator $\hat{\Box}$ on G, then reduce matters to the study of a "lifted" operator $\tilde X^2 + \tilde Y^2$ which may be closely modeled by the invariant operator $\hat{\Box}$ on G, and finally use the inverse of the latter to construct a parametrix for $\tilde X^2 + \tilde Y^2$.

We apply their method to $\Box + |D_\gamma|$. The first of their steps is replaced by our fourth, and their second goes through virtually unaltered. Their third is replaced by our fifth, but here serious complications arise because the convolution kernel for the model operator $\hat A$ on G has only a small finite number of derivatives away from 0, a reflection of the singularity of the convolution kernel for $|[\hat X, \hat Y]|$. We find it necessary to introduce a rather complicated technical modification of their method. The details will appear in [**C3**].

I am grateful to D. Jerison and E.M. Stein for pointing out an error in the statement of Theorem H in the original manuscript.

REFERENCES

[**BS**] L. Boutet de Monvel and J. Sjöstrand, *Sur la singularité des noyaux de Bergman et Szegö*, Soc. Mat. de France Asterisque **34-35** (1976), 123-164.

[**C1**] M. Christ, *Hilbert transforms along curves, I. Nilpotent groups*, Annals of Math. **122** (1985), 575–596.

[**C2**] ________, *On the regularity of inverses of singular integral operators*, to appear, Duke Math. J..

[**C3**] ________, *Regularity properties of the $\bar{\partial}_b$ equation on weakly pseudoconvex CR manifolds of dimension 3*, preprint October, 1987, submitted.

[**C4**] ________, *On the regularity of inverses of singular integral operators, II*, unpublished manuscript, July 1987.

[**C5**] ________, *Pointwise estimates for the relative fundamental solution of $\bar{\partial}_b$*, preprint September 15, 1987.

[**C6**] ________, *On the regularity of inverses of singular integral operators, III*, in preparation.

[**CG**] M. Christ and D. Geller, *Singular integral characterizations of Hardy spaces on homogeneous groups*, Duke Math. J. **51** (1984), 547–598.

[**CNSW**] M. Christ, A. Nagel, E.M. Stein and S. Wainger, in preparation.

[**CW**] R.R. Coifman and G. Weiss, "Analyse harmonique non–commutative sur certains espaces homogènes," Lecture Notes in Math. no. 242, Springer–Verlag, New York, 1971.

[**F**] C. Fefferman, *The Bergman kernel and biholomorphic mappings of pseudoconvex domains*, Invent. Math. **26** (1974), 1–65.

[**FK**] C. Fefferman and J.J. Kohn, *Hölder estimates on domains of complex dimension two and on three dimensional CR manifolds*, preprint.

[**Fo**] G.B. Folland, *Subelliptic estimates and function spaces on nilpotent groups*, Arkiv för Mat. **13** (1965), 161–207.

[**FS**] G.B. Folland and E.M. Stein, *Estimates for the $\bar{\partial}_b$ complex and analysis on the Heisenberg group*, Comm. Pure and Appl. Math. **27** (1974), 429–522.

[**H**] L. Hörmander, *Hypoelliptic second order differential equations*, Acta Math. **119** (1967), 147–171.

[**K1**] J.J. Kohn, *The range of the tangential Cauchy–Riemann operator*, Duke Math. J. **53** (1986), 525–545.

[**K2**] ________, *Estimates for $\bar{\partial}_b$ on pseudoconvex CR manifolds*, Proc. Symposia Pure Math. **43** (1985), 207–217.

[**K3**] ________, *Microlocal analysis on pseudoconvex domains and CR manifolds*, in preparation.

[**KS**] A. Knapp and E.M. Stein, *Intertwining operators for semisimple groups*, Annals of Math. **93** (1971), 489–578.

[**L**] H. Lewy, *An example of a smooth linear partial differential equation without solution*, Annals of Math. **66** (1957), 155–158.

[**M**] M. Machedon, *Estimates for the parametrix of the Kohn Laplacian on certain weakly pseudoconvex domains*, preprint.

[**N**] A. Nagel, *Vector fields and nonisotropic metrics*, in "Beijing Lectures in Harmonic Analysis," Princeton University Press, Princeton, N.J., 1986.

[**NRSW**] A. Nagel, J.P. Rosay, S. Wainger and E.M. Stein, *Estimates for the Bergman and Szegö kernels in certain weakly pseudoconvex domains*, announcement, preprint.

[**NS**] A. Nagel and E.M. Stein. personal communication.

[**NSW**] A. Nagel, E.M. Stein and S. Wainger, *Balls and metrics defined by vector fields I: Basic properties*, Acta Math. **155** (1985), 103–147.

[**RS**] L.P. Rothschild and E.M. Stein, *Hypoelliptic differential operators and nilpotent groups*, Acta Math. **137** (1976), 247–320.

SINGULAR INTEGRALS ON SURFACES

Guy David
Centre de Mathématiques
Ecole Polytechnique
91128 Palaiseau Cedex, France

In this lecture, I would like to describe a few classes of hypersurfaces S in R^{n+1}, $n \geq 2$, for which reasonable singular integrals--like the double-layer potential--define operators that are bounded on $L^2(S)$. Since the results I am about to mention are generalizations of one-dimensional results, I will first spend some time recalling a few facts about curves in the plane. Before that, I would like to say how much I appreciate the El Escorial meetings, and also thank J. L. Fernandez and S. Semmes, who helped prepare this lecture.

I. Singular integrals on curves

Let Γ be a rectifiable curve in the complex plane, and assume for simplicity that it is orientable and goes through ∞. Define the Cauchy operator on Γ by

$$C_\Gamma f(z) = \text{p.v.} \int_\Gamma \frac{1}{z-w} f(w)dw \quad \text{for suitable } f, \tag{1}$$

where dw is the complex-valued length measure on Γ.

Let us ask ourselves two questions about C_Γ. The first one is simply: When does C_Γ extend to a bounded operator on $L^2(\Gamma,|dw|)$? There is a rather simple geometric answer: if and only if Γ is "regular", i.e. iff there is a constant $C \geq 0$ such that, for all balls B in the plane, the total length of $\Gamma \cap B$ is less than C times the radius of B (see [5]).

Recall that Lipschitz graphs, or chord-arc curves (see Definition 1 below), are special cases of regular curves; however regular curves need not be Jordan, and may have cusps.

In order to state our second question, let us assume that Γ is also Jordan. Let Ω_1 and Ω_2 be the two connected components of $\mathbb{C}\setminus\Gamma$ (with, say, Ω_1 on the right of Γ) and let $H^2(\Omega_i)$, $i = 1,2$, denote the Hardy space of traces of analytic functions: $H^2(\Omega_i)$ is the closure, in $L^2(\Gamma)$, of the set of rational functions in $L^2(\Gamma)$ which have their poles away from Ω_i. Note that C_Γ is L^2-bounded iff $L^2(\Gamma)$ can be written as the direct sum

$$L^2(\Gamma) = H^2(\Omega_1) + H^2(\Omega_2) \tag{2}$$

This equivalence is, by the way, easy to prove because the projection from L^2 onto $H^2(\Omega_1)$ with kernel $H^2(\Omega_2)$ has a simple expression in terms of C_Γ. The direct sum (2) is never orthogonal, except when Γ is a straight line. Our second question is: In which sense can one say that, if the sum (2) is almost orthogonal, then Γ has to look like a straight line? More precisely, given a small $\varepsilon > 0$, we would like to know for which curves it is true that

$$\begin{cases} \text{the decomposition (2) holds, and also} \\ |\langle f_1, f_2 \rangle_{L^2(\Gamma)}| \le \varepsilon \|f_1\|_{L^2(\Gamma)} \|f_2\|_{L^2(\Gamma)} \\ \text{whenever } f_1 \in H^2(\Omega_1) \text{ and } f_2 \in H^2(\Omega_2). \end{cases} \tag{O_ε}$$

In terms of Cauchy integrals, (O_ε) means that C_Γ is almost anti-self-adjoint, i.e. that $\|C_\Gamma + C_\Gamma^*\|$ is small. The answer relies on the notion of chord-arc curves.

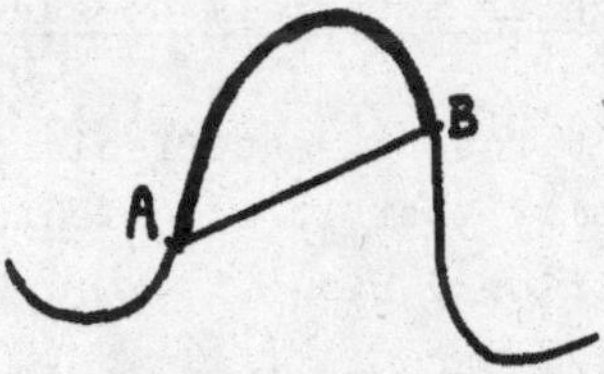

Definition 1. The curve Γ is chord-arc with constant $\le C$ if, for all points A, B in Γ, the length of the portion of Γ that joins A and B is $\le C|B-A|$.

Note that Γ is chord-arc iff it is the image of $\mathbb{R}$ by a bilipschitz map $z: \mathbb{R} \to \mathbb{C}$ [i.e. a map z such that $C^{-1}|u-v| \le |z(u)-z(v)| \le C|u-v|$ for some $C > 0$]. It is even true, in this dimension, that z can be extended into a bilipschitz homeomorphism of $\mathbb{C}$. Lipschitz graphs, logarithmic spirals [in polar coordinates, $\theta = K \log|r|$] are typical examples of chord-arc curves; when K tends to 0, the constant C of Definition 1 tends to 1.

The main part of the answer to our second question is due to Coifman and Meyer [2]: for each $\varepsilon > 0$, there is a $\delta > 0$ such that, if Γ is chord-arc with constant $\le 1+\delta$, then (O_ε) is satisfied. This can be seen by checking the speed at which the two Hardy spaces rotate when Γ moves in a neighborhood of a straight line, and also it can be seen as a consequence of the following more precise result of Coifman-Meyer. If $s \to z(s)$ is a parametrization of Γ by arclength, it is easily proved that Γ is chord-arc with small constant [one should probably say (1 + small) constant] iff $\|z'\|_{BMO}$ is small. One can even show that this is equivalent to $\|\log z'\|_{BMO}$ being small (when the "natural" determination of the log is taken). Coifman and Meyer proved that the mapping $\log z' \to C_\Gamma$, from a neighborhood of 0 in BMO to bounded operators on $L^2(\mathbb{R})$, is analytic near 0. One can easily deduce from this that $\|C_\Gamma + C_\Gamma^*\|$ tends to 0 with $\|\log z'\|$, and this gives the result.

There is a converse: for each $\delta > 0$, there exists $\varepsilon > 0$ such that (O_ε) implies that Γ is chord-arc, with constant $\leq 1+\delta$ (see [4]). This part is easier; the idea is that if $\|z'\|_{BMO}$ is large, then the kernel of $C_\Gamma + C_\Gamma^*$ is large somewhere, which is then enough to conclude that the operator is not small.

A few words can be said about the proofs of the direct results. To prove that C_Γ is bounded, one often starts from Calderón's theorem [C_Γ is bounded when Γ is the graph of a lipschitz function with small enough lipschitz norm]. One then uses good λ inequalities, together with the existence of so-called "Calderón-Zygmund decompositions" of curves [something like this: for each curve Γ of the type one wants to treat and for each interval I of Γ, there is a curve Γ_I of a better type--a lipschitz graph with small constant, for instance--that coincides with Γ on one-tenth of the length of I]. It is also possible to prove the boundedness of C_Γ for any chord-arc curve directly from the Tb-theorem [7] (which is, unfortunately, a little harder to establish).

II. Chord-arc surfaces with small constants [8]

In order to give a sense to our two questions in higher dimensions, let us recall a few facts about Clifford algebras (for more precisions on this, see [1]). For each integer $n \geq 1$, one can define a Clifford algebra. It is the algebra over $\mathbb{R}$ generated by $(n+1)$ linearly independent elements $e_0, e_1, \ldots, e_n$, where e_0 is a unit, and with the only extra relations $e_i^2 = -1$ for $i = 1, \ldots, n$ and $e_i e_j = -e_j e_i$ for $1 \leq i < j \leq n$. When $n = 1$, one obtains the complex numbers, and when $n = 2$, the quaternions. The elements of the Clifford algebra that are in the linear span of $e_0, e_1, \ldots, e_n$ are called Clifford vectors; we identify $\mathbb{R}^{n+1}$ with the set of Clifford vectors in the obvious way such that $\mathbb{R}^n$ is identified with the span of $e_1, \ldots, e_n$. One of the nice properties of Clifford vectors is that they are invertible as soon as they are not zero; however, the product of two vectors is not always a vector. The analogue of $\frac{\partial}{\partial \bar{z}}$ in this context is the Dirac operator $\mathcal{D} = \sum_{i=0}^{n} e_i \frac{\partial}{\partial x_i}$; one says that the Clifford-vectors-valued function f is Clifford-analytic if it satisfies $\mathcal{D}f = 0$ [this is a large system of equations, which reduces to the

Cauchy-Riemann equations when $n = 1$; also note that Clifford-analytic functions are harmonic--and thus C^∞--because $\mathcal{D}^*\mathcal{D} = (n+1)\Delta$]. Even when $n > 1$, there is a Cauchy kernel which allows one to recover the values of a Clifford-analytic function in the neighborhood of a domain from its boundary values; it is given by $K(z-w)$, where $K(w) = \frac{1}{\omega_n} \frac{w^*}{|w|^{n+1}}$, where ω_n is the surface of the unit sphere in $\mathbb{R}^{n+1}$, and $w^* = w_0 - \sum_{i=1}^{n} w_i e_i$ when $w = w_0 + \sum_{i=1}^{n} w_i e_i$. Note that K has the same homogeneity as the double layer potential (which is, incidentally, one of its components).

Let us come to singular integrals on hypersurfaces of $\mathbb{R}^{n+1}$. Let S be a n-dimensional submanifold of $\mathbb{R}^{n+1}$, and assume that it is oriented, goes through ∞, and of class C^1 (including at ∞, with the convention that $\mathbb{R}^{n+1} \cup \{\infty\} \simeq S^{n+1}$. [We shall not use the smoothless in a quantitative manner, but it is very convenient to use Stokes' theorem.] Call Ω_1 and Ω_2 the two connected components of $\mathbb{R}^{n+1} \setminus S$; one can define Hardy spaces $H^2(\Omega_1)$ and $H^2(\Omega_2)$ of traces of Clifford-analytic functions. Also, the Cauchy operator on S is defined by

$$C_S f(x) = \int_S K(y-x) n(y) f(y) d\sigma(y) \, , \tag{1'}$$

where K is as above, $n(y)$ is the unit normal to S at y pointing away from Ω_1, $d\sigma(y)$ is the surface measure on S, and f is a vector-valued function; naturally, all the products are taken in the Clifford algebra. As in dimension 1, our first question (when is C_S bounded on $L^2(S,d\sigma)$?) is equivalent to: "when does the decomposition

$$L^2(S) = H^2(\Omega_1) + H^2(\Omega_2) \quad \text{hold?"} \tag{2'}$$

Since this question turns out to be harder to answer, let us start with the second one: when is the decomposition (2') almost orthogonal in the sense of (O_ε)? Again, this decomposition (provided it exists) is never orthogonal, unless S is a hyperplane; the right notion of "looking like a hyperplane" is given in the next definition.

<u>Definition 2</u>. We will say that S is a chord-arc surface with small constant (in short a cassc) if, for some small $\delta > 0$,

the unit normal $n(y)$ is in $BMO(S,d\sigma)$ [the balls are still the Euclidean balls] with $\|n\|_{BMO} \le \delta$; (3)

for all $x \in S$ and $r > 0$, and all $y \in S \cap B(x,r)$, $|\langle y-x, n(x,r)\rangle| \le \delta r$, where $n(x,r)$ is the mean value of n on $S \cap B(x,r)$. (4)

Note that, since δ is small, $n(x,r)$ stays away from zero, and so (4) says that y is between two parallel hyperplanes whose distance is $\leq 4\delta r$. When $n = 1$, (3) is another way of saying that S is a chord-arc curve with small constant. We are not completely sure that (4) does not follow from (3), but it is quite unlikely.

Theorem (S. Semmes). There is a constant $k > 0$ such that

-- for all $0 < \varepsilon < 1$ and all casse S satisfying (3) and (4) with $\delta = k\varepsilon$, (O_ε) is true;
-- for all δ small enough and all surfaces S (with the above qualitative assumptions) such that (O_ε) is true with $\varepsilon = k\delta$, (3) and (4) hold.

So (3) and (4) are one way to measure the proximity of S to hyperplanes that answers our question. They imply a certain number of other properties, like the following:

for all $x \in S$ and $r > 0$, $\left|\frac{\sigma(S \cap B(x,r))}{\mu_n r^n} - 1\right| \leq \delta'$, where μ_n is the mass of the unit ball in $\mathbb{R}^n$; (5)

if $D(u,v)$ denotes the geodesic distance on S between two points u, v of S, then $|u-v| \leq D(u,v) \leq (1+\delta'')|u-v|$; (6)

for all $x \in S$ and $r > 0$, there is a surface $\tilde{S}$, which is the image by a direct isometry of a lipschitz graph with constant $\leq \delta'''$, and such that $\sigma(S \cap B(x,r) \setminus \tilde{S}) \leq \delta''' r^n$. (7)

Note how similar (5) and (6) are to the one-dimensional chord-arc condition; it is also true that (5) and (6), with small enough δ' and δ'', imply (3) and (4) [8].

Let us say a few words of the proof. The first assertion is obtained in a manner which is very similar to the proofs of boundedness of C_Γ. One uses (7) together with good λ estimates. For the second assertion, the idea is still, roughly speaking, that if $C_S + C_S^*$ has a small norm, then its kernel should be small everywhere, which should give geometric information on S. This part is made more tricky by the fact that one has, to begin with, very little information on the surface S [for instance, even when $n = 2$, the fact that S is homeomorphic to $\mathbb{R}^n$ requires a proof].

III. A sufficient condition [8]

Let us come back to the question of the boundedness of C_S on $L^2(S,d\sigma)$. A necessary condition is that

$$\text{for all } x \in S \text{ and } r > 0,\ \sigma(S \cap B(x,r)) \leq Cr^n; \tag{8}$$

on the other hand, by analogy with the case of curves and because it implies that S, with the measure $d\sigma$ and the Euclidean distance, is a space of homogeneous type, it seems natural to ask for

$$\text{for all } x \in S \text{ and } r > 0,\ \sigma(S \cap B(x,r)) \geq \frac{1}{C}r^n. \tag{9}$$

Unfortunately, (8) and (9) alone do not imply that all the reasonable singular integrals give L^2-bounded operators. This is connected with the fact that one can find measures σ in $\mathbb{R}^2 = \mathbb{C}$, satisfying (8) and (9) with $n = 1$, but such that the Cauchy kernel does not define a bounded operator on $L^2(d\sigma)$. The following picture gives a rough idea of how to transform the support of such a measure [the product by itself of a $1/4^{th}$ Cantor set] in a two-dimensional surface in $\mathbb{R}^3$ which is a counterexample:

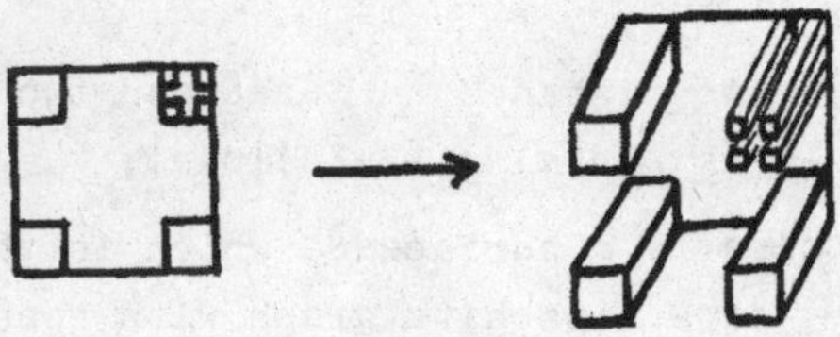

So, one looks for small extra conditions on S that would imply the boundedness of C_S. Here is one.

<u>Theorem</u> (S. Semmes [8]). Let S be a connected, orientable embedded submanifold of dimension n in $\mathbb{R}^{n+1}$, which is of class C^1 (including at ∞). Suppose that there is a constant $C \geq 0$ such that (8) and (9) are satisfied, as well as

for all $x \in S$ and $r > 0$, there exist two balls B_1 and B_2 of radius r, that are contained in two different connected components of $\mathbb{R}^{n+1} \setminus S$, and such that $\mathrm{dist}(x,B_i) \leq Cr$ for $i = 1$ and 2. (10)

Then C_S is bounded on $L^2(S,d\sigma)$, with an estimate on the norm which depends only on the constant C.

The proof of this theorem is a direct application of (the most complicated version of) the Tb-theorem [7]: conditions (8) and (9) imply that S is a space of homogeneous type (as in [3]); the Tb-theorem is then applied, in this context, to the Clifford-vector-valued function $b(y) = n(y)$ (the unit normal), and (10) is used to prove (with the aid of Stokes' theorem) that n is "pseudoaccretive." A certain number of extra integrations by parts allow S. Semmes to replace the Cauchy-Clifford kernel $K(w-z)$ by various other kernels; however it is not known whether any C^∞ kernel which is homogeneous of degree $-n$ and odd gives a bounded operator under the hypotheses of the theorem. It would certainly be interesting to have a different proof of this theorem, for instance using the standard "good λ" techniques; such a proof, if it existed, would probably be less rigid, apply to all reasonable kernels, and provide a better understanding of the geometry.

Examples of surfaces satisfying (8), (9) and (10) include lipschitz graphs, images of $\mathbb{R}^n$ by a bilipschitz function $z: \mathbb{R}^n \to \mathbb{R}^{n+1}$, cassc; also, the topology of S can be very complicated: S can have many handles, for instance. In dimension 1, the notion is more general than "chord-arc curves", and less general than "regular curves". See a few examples in the pictures:

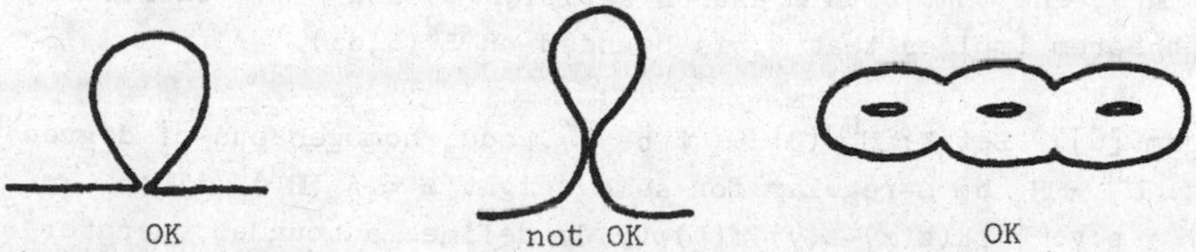

OK (dimension n = 1) — not OK (n = 1) — OK (n = 2)

IV. Regular surfaces

When one tries to generalize the results of section 1 to surfaces, the main difficulty which one encounters is that there is no canonical way of parametrizing a surface, even locally. This difficulty can be overcome in the case of a cassc because it looks very much like hyperplanes; also, in section 3, it was possible to obtain a result by using a theorem of global nature (the drawback is that the proof becomes mysterious). In this section, we shall deal with the problem more brutally: we shall simply assume that S is given to us together with a nice parametrization. With this assumption, it is again possible to carry out the same program as in one dimension: use good λ estimates, and obtain a L^2-boundedness result.

Definition 3. Let $N \geq n \geq 1$ be integers, and let ω be a weight in the class $A_\infty(\mathbb{R}^n)$ of Muckenhoupt. A function $z: \mathbb{R}^n \to \mathbb{R}^N$ is said to be "ω-regular" if, for some constant $C \geq 0$,

$|\nabla z(x)| \leq C\omega(x)^{1/n}$ (in the sense of distributions), and (11)

for all $u \in \mathbb{R}^N$ and $r > 0$, (12)

$$\omega(\{x \in \mathbb{R}^n;\ |z(x)-u| \leq r\}) \leq Cr^n .$$

We shall also say that $S = z(\mathbb{R}^n)$ is an ω-regular surface; if $\omega = 1$, we shall simply say that z and S are regular.

The special case $\omega = 1$ already gives a good idea of what these surfaces look like [in fact, the author is not completely sure yet that there are ω-regular surfaces that cannot be reparametrized as regular surfaces]. Lipschitz graphs, bilipschitz images of $\mathbb{R}^n$ in $\mathbb{R}^N$ are regular surfaces; other trivial examples can be obtained by taking the product of a regular curve by $\mathbb{R}^{n-1}$. One proves that, if z is regular, then the image by z of the measure $\omega\, dx$ is equivalent to (i.e. is more than C^{-1} times and less than C times) the surface measure $d\sigma$ on $S = z(\mathbb{R}^n)$, and that both measures satisfy (8) and (9). Therefore, the next theorem implies that C_S is bounded on $L^2(S, d\sigma)$.

Theorem [6]. Let $K: \mathbb{R}^N - \{0\} \to \mathbb{C}$ be C^∞, odd, homogeneous of degree $-n$. Let $z: \mathbb{R}^n \to \mathbb{R}^N$ be ω-regular for some weight $\omega \in A_\infty(\mathbb{R}^n)$. Then $Tf(x) = \text{p.v.} \int_{\mathbb{R}^n} K(z(x)-z(y))f(y)\omega(y)dy$ defines a bounded operator on $L^2(\mathbb{R}, \omega dx)$.

If one already knows that z satisfies (11), then (12) is necessary if one wants all the functions K to give a bounded operator, but the strong assumption is the existence of z: for instance the surfaces of section 4 can have so many handles that they cannot be covered by regular surfaces.

V. A few questions

It should be clear, at this point of the lecture, that there are many aspects of the problem which we do not understand. Finding necessary and sufficient conditions for C_S to be bounded seems out of the question for the moment (it would probably imply knowing for which measures σ on $\mathbb{C}$ the Cauchy integral defines a bounded operator on $L^2(\sigma)$). One can expect some progress, for instance, from the methods mentioned in P. Jones' lecture, however. There are much simpler questions, though, that we cannot answer yet. For instance, is it true that every cassc (with δ small enough) can be parametrized in an ω-regular manner for some weight ω? This was proved by S. Semmes for two-dimensional surfaces (using the existence of isothermic coordinates), but in higher dimension, we only know that S is homeomorphic to $\mathbb{R}^n$, with estimates that are not quite good enough [8]. When $n = 2$, we do not know whether S has a regular parametrization (i.e., with $\omega = 1$).

Connected questions concern the absolute continuity of the harmonic measure associated to one of the components Ω of $\mathbb{R}^{n+1}\backslash S$, with respect to $d\sigma$. For instance, is it sufficient that Ω be a N.T.A. domain, and S satisfy (8) and (9)? [We probably can do it in the case of a bilipschitz image of $\mathbb{R}^n$.] Also, when is it true that if f is in the Hardy space $H^1(\Omega)$ of traces of Clifford-analytic functions, then $\|Mf\|_{L^1} \leq C\|f\|_{L^1}$, where Mf is the non-tangential maximal function? When does one have decompositions like $BMO(S) = L^\infty + C_S L^\infty$, or $BMO(S) = L^\infty + \sum_j R_j(BMO)$ for some reasonable set of singular integral operators R_j? The two last questions should be closely related to harmonic measure estimates.

References

[1] F. Brackx, R. Delanghe, F. Sommer, Clifford Analysis, Pitman, 1982.

[2] R. Coifman, Y. Meyer, Une généralisation du théorème de Calderón sur l'intégrale de Cauchy, Proc. Conf. on Fourier Analysis at El Escorial, Spain (1979), edited by M. de Guzman and I. Peral.

[3] R. Coifman, G. Weiss, Analyse harmonique non-commutative sur certains espaces homogènes, Lect. Notes in Math. 242, Springer-Verlag, 1971.

[4] G. David, Courbes corde-arc et espaces de Hardy généralisés, Ann. Inst. Fourier, Grenoble 32, 3 (1982), 227-239.

[5] G. David, Opérateurs intégraux singuliers sur certaines courbes du plan complexe, Ann. Scient. Ec. Norm. Sup. 17 (1984), 157-189.

[6] G. David, Opérateurs d'intégrale singulière sur les surfaces régulières, preprint.

[7] G. David, J.-L. Journé, S. Semmes, Opérateurs de Calderón Zygmund, fonctions para-accrétives et interpolation, Revista Matematica Ibero-americana 1, 4 (1985), 1-56. (The preliminary preprint in English contains more information on pseudoaccretive functions.)

[8] S. Semmes, personal communications.

The T1 Theorem for Triebel–Lizorkin spaces.

by

M. Frazier, Y.–S. Han, B. Jawerth, G. Weiss

§1. **Introduction.** Let $\mathcal{D} = \mathcal{D}(\mathbb{R}^n)$ denote the space of Schwartz test functions $\{\varphi \in C^\infty(\mathbb{R}^n)$: Supp φ is compact$\}$ and $\mathcal{D}'$ the space of Schwartz distributions (the dual of $\mathcal{D}$). Suppose $T: \mathcal{D} \to \mathcal{D}'$ is a continuous linear operator. The Schwartz kernel theorem (see [6]) tells us that there exists a distribution $K \in \mathcal{D}'(\mathbb{R}^n \times \mathbb{R}^n)$ such that

$$\langle T\varphi, \psi \rangle = K(\psi \otimes \varphi) \tag{1.1}$$

for all $\varphi, \psi \in \mathcal{D}$. K is called the *kernel* of T. We say that K is a *Calderón–Zygmund kernel* if it is "represented" by a continuous function on $\{(x,y) \in \mathbb{R}^n \times \mathbb{R}^n: x \neq y\}$, which we also denote by K, satisfying:

$$\begin{aligned} &(\mathrm{i}) \quad |K(x,y)| \le C|x-y|^{-n} \\ &(\mathrm{ii}) \quad |K(x',y) - K(x,y)| + |K(y,x') - K(y,x)| \le \\ &\qquad C|x-x'|^{\epsilon}|x-y|^{-(n+\epsilon)}, \text{ provided } 2|x'-x| \le |x-y|, \end{aligned} \tag{1.2}$$

for an ϵ in $(0,1]$; moreover,

$$\langle T\varphi, \psi \rangle = \iint_{\mathbb{R}^n \times \mathbb{R}^n} K(x,y)\varphi(y)\psi(x)\,dx\,dy \tag{1.3}$$

whenever $\varphi, \psi \in \mathcal{D}$ and (Supp φ) $\cap$ (Supp ψ) $= \phi$. (Throughout this paper the symbol C will denote different absolute constants). When T is associated with such a kernel we write $T \in CZK$ (or $T \in CZK(\epsilon)$ if we want to emphasize that the "smoothness" condition (1.2) (ii) is "Lipschitz of order ϵ").

If $T \in CZK$ and has a bounded extension as an operator from $L^2(\mathbb{R}^n)$ into $L^2(\mathbb{R}^n)$ it is called a *Calderón–Zygmund operator* and we write $T \in CZO$.

A basic problem is to find conditions that assure us that if $T \in CZK$ then $T \in CZO$. The most significant result in this direction is a theorem of G. David and J.–L. Journé [3]. In order to

This research was supported by NSF Grants DMS–8604528 and MCS–8200884.

state their result we need to introduce some concepts that will be necessary for our development as well. The first of these concepts is the ***Weak Boundedness Property*** (WBP): A continuous linear operator $T:\mathcal{D}\longrightarrow\mathcal{D}'$ is said to satisfy this property iff for all $\varphi,\psi\in\mathcal{D}$ with supports having diameter at most $t>0$,

$$|\langle T\varphi,\psi\rangle| \leq Ct^n(\|\varphi\|_\infty + t\|\nabla\varphi\|_\infty)(\|\psi\|_\infty + t\|\nabla\psi\|_\infty). \tag{1.4}$$

It is clear that if T is a bounded operator on $L^2(\mathbb{R}^n)$ it then satisfies the weak boundedness property (which we shall denote by writing $T\in WBP$).

To a continuous linear operator $T:\mathcal{D}\longrightarrow\mathcal{D}'$ we can always associate a transpose $T^*:\mathcal{D}\longrightarrow\mathcal{D}'$ defined by putting $\langle T^*\psi,\varphi\rangle = \langle T\varphi,\psi\rangle$ for all $\varphi,\psi\in\mathcal{D}$. Suppose $\psi\in\mathcal{D}$ and $x\notin \mathrm{Supp}\,\psi$. Then

$$\theta(x) = \int_{\mathbb{R}^n} K(y,x)\psi(y)\,dy$$

is well defined and, by (1.3) and the definition of T^*, we have

$$\langle T^*\psi,\varphi\rangle = \int_{\mathbb{R}^n}\theta(x)\varphi(x)\,dx$$

whenever $(\mathrm{Supp}\,\varphi)\cap(\mathrm{Supp}\,\psi)=\phi$. Thus, we write

$$\theta(x) = (T^*\psi)(x) = \int_{\mathbb{R}^n} K(y,x)\psi(y)\,dy$$

for $x\notin\mathrm{Supp}\,\psi$.

Suppose $y_o\in\mathrm{Supp}\,\psi$ and $|x-y_o|\geq 2d$, where d is the diameter of the support of ψ. Then, for $y\in\mathrm{Supp}\,\psi$ we have $|y-y_o|\leq d\leq|x-y_o|/2$ and, from (1.2) (ii),

$$|K(y,x)-K(y_o,x)|\leq C|y-y_o|^\epsilon|x-y_o|^{-(n+\epsilon)}.$$

If, in addition, $\int_{\mathbb{R}^n}\psi(x)\,dx=0$ then

$$|(T^*\psi)(x)| = \left|\int_{\mathbb{R}^n}[K(y,x)-K(y_o,x)]\psi(y)\,dy\right| \leq$$

$$C\int_{\mathbb{R}^n}|y-y_o|^\epsilon|x-y_o|^{-(n+\epsilon)}|\psi(y)|\,dy \leq Cd^\epsilon\|\psi\|_1|x-y_o|^{-(n+\epsilon)}.$$

That is,

$$|(T^*\psi)(x)|\leq C|x-y_o|^{-(n+\epsilon)} = C(T,\psi)|x-y_o|^{-(n+\epsilon)}. \tag{1.5}$$

This allows us to define Tf, for f a bounded C^∞ function on $\mathbb{R}^n$, as a linear functional on the space $\mathcal{D}_o = \{\psi\in\mathcal{D}: \int_{\mathbb{R}^n}\psi(x)\,dx=0\}$: Choose $\psi\in\mathcal{D}_o$ and let $1=\chi_o+\chi_1$, where $\chi_o\in\mathcal{D}$ and $\chi_o(y)$

$= 1$ if $|y - y_0| \le 2d$ (where, again, $y_0 \in \text{Supp}\,\psi$ and d is the diameter of the support of ψ). Then, from (1.5) we see that $\int_{\mathbb{R}^n} f(x)\chi_1(x)(T^*\psi)(x)\,dx$ is well defined and we write

$$<T^*\psi, f\chi_1> = \int_{\mathbb{R}^n} f(x)\chi_1(x)(T^*\psi)(x)\,dx$$

and put

$$<Tf,\psi> \equiv <T(f\chi_0),\psi> + <T^*\psi, f\chi_1> \,{}^{(2)}. \tag{1.6}$$

It is an easy exercise to show that this definition of $<Tf,\psi>$ is independent of the choice of the pair χ_0, χ_1 satisfying the above properties.

We can now state the result of David and Journé:

Theorem (The T1 theorem of David and Journé). ***Suppose*** **T ∈ CZK.** ***Then*** **T ∈ CZO** ***iff*** **(a) T1 ∈ BMO, (b) T^*1 ∈ BMO** ***and*** **(c) T ∈ WBP.**

This theorem can be reduced to a weaker one which asserts that if $T \in CZK \cap WBP$ and $T1 = T^*1 = 0$ (the zero element in BMO) then $T \in CZO$. In this paper we shall present a rather direct proof of this (reduced) result. Briefly our argument is as follows: We first show that if $T \in CZK \cap WBP$ and $T1 = 0$, then T is a bounded operator on the Triebel–Lizorkin spaces $\dot{F}_p^{\alpha,q}(\mathbb{R}^n)$ (which will be defined below) for $1 < p,q < \infty$ and appropriate $\alpha \in (0,1)$. If $T^*1 = 0$ as well, we can then apply this result to T^*, which gives us the boundedness of T on the space $\dot{F}_{p'}^{-\alpha,q'}(\mathbb{R}^n)$ dual to $\dot{F}_p^{\alpha,q}$ $(\frac{1}{p} + \frac{1}{p'} = 1 = \frac{1}{q} + \frac{1}{q'})$. But $L^2(\mathbb{R}^n)$ is an intermediate space (in the sense of interpolation theory) for this pair of spaces, $\dot{F}_p^{\alpha,q}$ and its dual. It follows, therefore, that T is bounded on $L^2(\mathbb{R}^n)$ and, thus, $T \in CZO$.

A corresponding result for Besov spaces $\dot{B}_p^{\alpha,q}$ was obtained by Lemarié [7]. If $\dot{F}_p^{\alpha,q}$ is replaced by $\dot{B}_p^{\alpha,q}$, Lemarié's result reads precisely as ours. We observe that his result is a consequence of ours. To see this we use the real interpolation theory result: $\dot{B}_p^{\alpha,q} =$

---

(2) This is consistent with the formal equalities $<Tf,\psi> = <Tf(\chi_0+\chi_1),\psi> = <T(f\chi_0),\psi> + <T(f\chi_1),\psi> = <T(f\chi_0,\psi> + <T^*\psi, f\chi_1>$.

$(\dot{B}^{\alpha_0,p_0}_{p_0},\dot{B}^{\alpha_1,p_1}_{p_1})_{\theta,q}$ where $\alpha = (1-\theta)\alpha_0+\theta\alpha_1$, $1/p = (1-\theta)(1/p_0)+\theta(1/p_1)$ with $0 < \theta < 1$. Our claim then follows from the identity $\dot{B}^{\alpha,p}_p = \dot{F}^{\alpha,p}_p$ (which is immediate from the definition of these spaces that we shall present below).

The definition of the Besov spaces that we shall use is the one given in Peetre's book [10]: Let $\varphi \in \mathscr{S}(\mathbb{R}^n)$ (the space of tempered test functions) satisfy Supp $\hat{\varphi} \subset \{\xi \in \mathbb{R}^n: 1/2 \le |\xi| \le 2\}$ and $|\hat{\varphi}(\xi)| \ge C > 0$ when $3/5 \le |\xi| \le 5/3$; put $\varphi_\nu(x) = 2^{\nu n}\varphi(2^\nu x)$. If $\alpha \in \mathbb{R}$ and $0 < p,q \le \infty$ then the space $\dot{B}^{\alpha,q}_p$ is the collection of all $f \in \mathscr{S}'/\mathscr{P}$ (the tempered distributions modulo polynomials) satisfying

$$\|f\|_{\dot{B}^{\alpha,q}_p} = \Big(\sum_{\nu\in\mathbb{Z}} (2^{\nu\alpha}\|\varphi_\nu * f\|_{L^p})^q\Big)^{1/q} < \infty.$$

The Triebel–Lizorkin spaces are defined in a similar way, for $0 < p < \infty$, except that the order of taking the ℓ^q and the L^p norms is reversed. More precisely, $\dot{F}^{\alpha,q}_p$ is the collection of all $f \in \mathscr{S}'/\mathscr{P}$ such that

$$\|f\|_{\dot{F}^{\alpha,q}_p} = \Big\|\Big(\sum_{\nu\in\mathbb{Z}} (2^{\nu\alpha}|\varphi_\nu * f|)^q\Big)^{1/q}\Big\|_{L^p} < \infty.$$

The fact that the Besov spaces and the Triebel–Lizorkin spaces are the same when $p = q$ follows immediately from these definitions.

Frazier and Jawerth [4] have shown that the elements in $\dot{F}^{\alpha,q}_p$ can be expressed in terms of certain building blocks called smooth atoms for $\dot{F}^{\alpha,q}_p$. In this paper we shall only consider the case $\alpha \in (0,1)$ and $p,q \ge 1$. In this case these atoms are functions $a_Q \in \mathscr{D}$ associated with the dyadic cubes $Q = Q_{\nu k} = \{x = (x_1,x_2,\dots,x_n) \in \mathbb{R}^n: k_i \le 2^\nu x_i \le k_i + 1,\ i = 1,\dots,n\}$, where $k \in \mathbb{Z}^n$ and $\nu \in \mathbb{Z}$, satisfying

(1.7) $$\operatorname{Supp} a_Q \subset 3Q,$$

(1.8) $$\int x^\gamma a_Q(x)\,dx = 0 \text{ if } |\gamma| \le N,$$

(1.9) $$|\partial^\gamma a_Q(x)| \le |Q|^{-(1/2)-(|\gamma|/n)} \text{ if } |\gamma| \le K,$$

where $K \geq 1$, $N \geq -1$, $\gamma = (\gamma_1,\dots,\gamma_n) \in \mathbb{Z}^n$ and $|\gamma| = \gamma_1 + \gamma_2 + \dots + \gamma_n$ (if $\gamma_i < 0$ for some i, (1.8) is void).

We call these functions (K,N) smooth atoms for $\dot{F}_p^{\alpha,q}$.

It is convenient to introduce the sequence space $\dot{f}_p^{\alpha,q}$ associated with the Triebel–Lizorkin space having the same indices: $s = \{s_Q\} \in \dot{f}_p^{\alpha,q}$ iff it is a sequence indexed by the dyadic cubes in $\mathbb{R}^n$ satisfying

$$\|s\|_{\dot{f}_p^{\alpha,q}} = \|(\Sigma_Q(|Q|^{-\alpha/n}|s_Q|\tilde{\chi}_Q)^q)^{1/q}\|_{L^p} < \infty,$$

where $\tilde{\chi}_Q = |Q|^{-1/2}\chi_Q$. Then we have the following (smooth) **atomic decomposition** for the Triebel–Lizorkin spaces (which we state here for general indices α, p, q):

Theorem (1.10). *Let $\alpha \in \mathbb{R}$, $0 < q \leq \infty$ and $0 < p < \infty$. If $f \in \dot{F}_p^{\alpha,q}$, there exists a family of (K,N) smooth atoms $\{a_Q\} \subset \mathcal{D}_0$ and a sequence of coefficients $s = \{s_Q\}$ such that $f = \Sigma_Q s_Q a_Q$ (in $\mathcal{S}'/\mathcal{P}$) and $\|s\|_{\dot{f}_p^{\alpha,q}} \leq C\|f\|_{\dot{F}_p^{\alpha,q}}$. Conversely, $\|\Sigma_Q s_Q a_Q\|_{\dot{F}_p^{\alpha,q}} \leq C\|s\|_{\dot{f}_p^{\alpha,q}}$ for any family of (K,N) smooth atoms $\{a_Q\}$, if K and N are sufficiently large, depending on α, p, and q.*

The atomic decomposition of the Hardy spaces H^p, $0 < p \leq 1$, has been used to study the behaviour of operators acting on these spaces. It is too much to hope that a $T \in CZO$ maps atoms into atoms; nevertheless, there are "building blocks" that are more general than atoms, called **molecules**, that are the images, under T, of atoms. This fact, togetherwith an **appropriate molecular decompostion theorem** can the be used to show that T is continuous on H^p (see [2] or [11]). It turns out that a similar situation is true for the Triebel–Lizorkin spaces

Let $\alpha \in (0,1)$, $p,q \geq 1$ and Q a dyadic cube of side length $\ell(Q)$. Then a function m_Q is a **smooth molecule** for $\dot{F}_p^{\alpha,q}$ if and only if there exists δ and M satisfying $\alpha < \delta \leq 1$, $n < M \leq n+1$, such that

(1.11) $$|m_Q(x)| \leq |Q|^{-1/2}(1+\ell(Q))^{-1}|x-x_Q|)^{-M}$$

and (1.12):

$$|m_Q(x)-m_Q(y)| \leq |Q|^{-1/2}(|x-y|/\ell(Q))^{\delta}\{(1+|x-x_Q|/\ell(Q))^{-M} + (1+|y-x_Q|/\ell(Q))^{-M}\},$$

where x_Q is the "lower left corner" of Q.

We shall also refer to such functions as (δ,M)–molecules. Frazier and Jawerth [5] show that there exists a molecular decomposition of the Triebel–Lizorkin spaces. In particular, we have:

Theorem (1.13). *If* $f = \Sigma_Q s_Q m_Q$, *where* $\{m_Q\}$ *is a family of* (δ,M) *smooth molecules for* $\dot{F}_p^{\alpha,q}$, *then*

$$\|f\|_{\dot{F}_p^{\alpha,q}} \le C\|\{s_Q\}\|_{\dot{f}_p^{\alpha,q}}.$$

The principal result of this paper can be stated as follows:

Theorem (1.14). *Suppose* $0 < \alpha < \epsilon \le 1$ *and* $1 \le p,q < \infty$. *If* $T \in CZK(\epsilon) \cap WBP$ *and* $T1 = 0$, *then* T *is bounded on* $\dot{F}_p^{\alpha,q}$.

It is clear from theorems (1.10) and (1.13) that this result follows immediately from:

Theorem (1.15). *Under the hypotheses of the last theorem there exists a constant* C, *depending only on* T, *such that, if* $\epsilon < 1$, CTa *is an* $(\epsilon,n+\epsilon)$ *molecule whenever* a *is a* $(1,0)$ *atom; if* $\epsilon = 1$ *then* CTa *is a* $(\beta,n+1)$ *molecule for* $0 < \alpha < \beta < 1$.

Meyer [9] obtains the more general result that T maps molecules into molecules. We only need the (simpler) fact that T maps atoms into molecules; hence, we will present a proof of Theorem (1.15) in section 2.

To describe a result in the converse direction, we first note that an operator T defined on $\mathcal{D}_0$ that maps each $(1,0)$ smooth atom to a $(\beta,n+\epsilon)$ molecule, $0 < \beta < \epsilon \le 1$, extends to a bounded operator $\tilde{T}$ on $\dot{F}_p^{\alpha,q}$ for $0 < \alpha < \beta$, $1 < p,q < \infty$, by theorems (1.10) and (1.13). We also note that for $\alpha > 0$ and $1 < p,q < \infty$, we have $\mathcal{D} \subset \dot{F}_p^{\alpha,q}$. This follows from theorem (1.13) since any element of $\mathcal{D}$ is , in fact, a multiple of a $(1,n+1)$ molecule. In section 3 we prove the following (converse) result:

Theorem (1.16). *Suppose* T *is a continuous linear operator from* $\mathcal{D}_0$ *into* $\mathcal{D}'$ *which maps each* (1,0) *smooth atom to a* $(\beta, n+\epsilon)$ *molecule,* $0 < \beta < \epsilon \le 1$. *Then the kernel of* $\tilde{T}\big|_{\mathcal{D}}$, K(x,y), *satisfies the following Calderón–Zygmund estimates:*

(1) $$|K(x,y)| \le C|x-y|^{-n},$$

(2) $$|K(x,y) - K(x',y)| \le C|x-x'|^{\beta}|x-y|^{-n-\beta}$$

provided $|x-y| \ge 2|x-x'|$.

§2. Proof of Theorem (1.15). Two basic consequences of the hypotheses $T \in CZK(\epsilon) \cap WBP$ and $T1 = 0$ are the following two lemmas that can be found in M. Meyer [8] and Y. Meyer [9]:

Lemma (2.1). *Suppose* T *satisfies these hypotheses and* $\varphi \in \mathcal{D}$ *satisfies* Supp $\varphi \subset Q$, *a cube in* $\mathbb{R}^n$. *Then*

$$|T\varphi|_\infty \le C(|\varphi|_\infty + |Q|^{1/n}|\nabla\varphi|_\infty),$$

where C *is a constant dependent only on* T.

Lemma (2.2). *Suppose* T *satisfies these hypotheses and* $\varphi \in \mathcal{D}$. *Let* $x_1 \ne x_2$ *belong to* $\mathbb{R}^n$, $\xi(x) = \xi_0((x-x_1)/|x_2-x_1|)$, $\eta = 1-\xi$, *where* ξ_0 *is a fixed function in* $\mathcal{D}$ *satisfying* $\xi_0(x) = 1$ *for* $|x| \le 10$ *and* $\xi_0(x) = 0$ *for* $|x| \ge 20$. *We then have*

$$\begin{aligned} T\varphi(x_2) - T\varphi(x_1) &= \int [K(x_2,y) - K(x_1,y)][\varphi(y) - \varphi(x_1)]\eta(y)\,dy \\ &\quad - \int_{y \ne x_1} K(x_1,y)[\varphi(y) - \varphi(x_1)]\xi(y)\,dy \\ &\quad + \int_{y \ne x_2} K(x_2,y)[\varphi(y) - \varphi(x_2)]\xi(y)\,dy \\ &\quad + [\varphi(x_2) - \varphi(x_1)]T\xi(x_2). \end{aligned}$$

We begin the proof of (1.15) by showing that Ta satisfies the molecular condition (1.11) whenever $a = a_Q$ is a (1,0) smooth atom supported in 3Q. More precisely we shall show that

(2.3) $$|Ta(x)| \le C|Q|^{-1/2}(1 + 2^{\mu}|x - x_Q|)^{-n-\epsilon},$$

where $\ell(Q) = 2^{-\mu}$ is the side length of the dyadic cube Q, and C is a constant that is independent of the choice of the atom a. It is easily checked that, because of the form of the estimates (1.2), we can reduce the problem to the case where Q has lower left corner x_Q at the origin.

Suppose, first, that $|x| \geq 6\sqrt{n}2^{-\mu}$. We then have

$$Ta(x) = \int K(x,y)a(y)\,dy = \int [K(x,y) - K(x,0)]a(y)\,dy.$$

Since $|y| \leq 2\sqrt{n}2^{-\mu}$, we have $|x-y| \geq |x|-|y| \geq 4\sqrt{n}2^{-\mu} \geq 2|y|$ and we can use (1.2) (ii) to obtain

$$|Ta(x)| \leq C \int_{|y| \leq 2\sqrt{n}2^{-\mu}} \frac{|y|^{\epsilon}}{|x-y|^{n+\epsilon}} |a(y)|\,dy$$

$$\leq C|x|^{-n-\epsilon} \int_{|y| \leq 2\sqrt{n}2^{-\mu}} |y|^{\epsilon}|a(y)|\,dy \leq C|Q|^{-1/2}(2^{\mu}|x|)^{-n-\epsilon}$$

and (2.3) holds for such x (we used (1.9), with $|\gamma| = 0$, to obtain the last inequality).

Suppose, now, that $|x| \leq 6\sqrt{n}2^{-\mu}$. We then use (2.1) and (1.9), with $|\gamma| = 0, 1$, to obtain

$$\|Ta\|_{\infty} \leq C(\|a\|_{\infty} + 2^{-\mu}\|\nabla a\|_{\infty}) \leq C|Q|^{-1/2},$$

which certainly implies (2.3) for $|x| \leq 6\sqrt{n}2^{-\mu}$ as well.

We now must show that Ta satisfies the molecular condition (1.12). More precisely, we shall show that

$$|Ta(x_2) - Ta(x_1)| \leq \tag{2.4}$$
$$C|Q|^{-1/2}(2^{\mu}|x_1-x_2|)^{\epsilon}\{(1+2^{\mu}|x_1|)^{-n-\epsilon} + (1+2^{\mu}|x_2|)^{-n-\epsilon}\}.$$

Let us first suppose that $|x_1-x_2| \geq 2^{-\mu}$. From (2.3) we have

$$|Ta(x_2) - Ta(x_1)| \leq |Ta(x_2)| + |Ta(x_1)| \leq$$
$$C|Q|^{-1/2}\{(1+2^{\mu}|x_1|)^{-n-\epsilon} + (1+2^{\mu}|x_2|)^{-n-\epsilon}\} \leq$$
$$C|Q|^{-1/2}(2^{\mu}|x_1-x_2|)^{\epsilon}\{(1+2^{\mu}|x_1|)^{-n-\epsilon} + (1+2^{\mu}|x_2|)^{-n-\epsilon}\}.$$

Thus, (2.4) is established when $|x_1-x_2| \geq 2^{-\mu}$.

Let us now suppose $|x_1-x_2| < 2^{-\mu}$. We consider several cases:

Case 1: $|x_1|, |x_2| \geq 10\sqrt{n}2^{-\mu}$. Then we have

$$|Ta(x_2) - Ta(x_1)| = |\int[K(x_2,y) - K(x_1,y)]a(y)\,dy|.$$

Since $|y| \leq 2\sqrt{n}\cdot 2^{-\mu}$, $|x_1-x_2| \leq 2^{-\mu}$ and $|x_1| \geq 10\sqrt{n}2^{-\mu}$ we must have $|x_1-y| \geq 2|x_1-x_2|$; hence,

$$|\int[K(x_2,y) - K(x_1,y)]a(y)\,dy| \leq C\int \frac{|x_1-x_2|^{\epsilon}}{|x_1-y|^{n+\epsilon}} |a(y)|\,dy \leq$$

$$C|Q|^{-1/2}|x_1-x_2|^{\epsilon}|x_1|^{-n-\epsilon} \int_{|y|\leq 2\sqrt{n}2^{-\mu}} dy \leq$$

$$C|Q|^{-1/2}(2^{\mu}|x_1-x_2|)^{\epsilon}(2^{\mu}|x_1|)^{-n-\epsilon} \leq$$

$$C|Q|^{-1/2}(2^{\mu}|x_1-x_2|)^{\epsilon}\{(1+2^{\mu}|x_1|)^{-n-\epsilon} + (1+2^{\mu}|x_2|)^{-n-\epsilon}\},$$

which is (2.4).

Case 2: $|x_1| < 10\sqrt{n}2^{-\mu}$, $|x_2| < 10\sqrt{n}2^{-\mu}$. Let $r = |x_1-x_2|$, $\varphi_r(u) = \varphi(u/r)$, where $\varphi \in \mathcal{D}$ with $\varphi(u) = 1$ if $|u| \leq 10$ and $\varphi(u) = 0$ if $|u| \geq 20$. Letting $\overline{\varphi} = 1-\varphi$ we can then apply lemma (2.2) to obtain the equality:

$$Ta(x_1) - Ta(x_2) = \tag{2.5}$$

$$\int[K(x_1,y)-K(x_2,y)](a(y)-a(x_1))\overline{\varphi}_r(y-x_1)\,dy + \int_{y\neq x_1} K(x_1,y)(a(y)-a(x_1))\varphi_r(y-x_1)\,dy$$

$$- \int_{y\neq x_2} K(x_2,y)(a(y)-a(x_2))\varphi_r(y-x_1)\,dy + (a(x_1)-a(x_2))T\varphi_r(x_2).$$

In order to estimate the last term in this sum we can use lemma (2.1) to obtain

$$\|T\varphi_r\|_{\infty} \leq C, \tag{2.6}$$

where C depends only on φ and not on r (observe that $\|\varphi_r\|_\infty = \|\varphi\|_\infty$, Supp $\varphi_r \subset \tilde{Q}$ with $|\tilde{Q}|^{1/n} \approx r$ and, thus, $|\tilde{Q}|^{1/n}\|\nabla\varphi_r\|_\infty \approx \|\nabla\varphi\|_\infty$). Hence, using (1.9) with $|\gamma| = 1$,

$$|(a(x_1)-a(x_2))T\varphi_r(x_2)| \le C|x_1-x_2||Q|^{-(1/2)-(1/n)}$$

$$=C|Q|^{-1/2}(2^\mu|x_1-x_2|) \le C|Q|^{-1/2}(2^\mu|x_1-x_2|)^\epsilon$$

$$\le C|Q|^{-1/2}(2^\mu|x_1-x_2|)^\epsilon\{(1+2^\mu|x_1|)^{-n-\epsilon} + (1+2^\mu|x_2|)^{-n-\epsilon}\}.$$

To obtain the last inequality we use the "Case 2 assumption" $2^\mu|x_j| < 10\sqrt{n}$, $j = 1, 2$; to obtain the second to last inequality we use $2^\mu|x_1-x_2| < 1$ and $0 < \epsilon \le 1$.

The estimation of the first summand in (2.5) is direct:

Let $\mathscr{R} = \{y \in \mathbb{R}^n\colon 10|x_1-x_2| \le |y-x_1| \le 11\sqrt{n}2^{-\mu}\}$ and $\mathscr{T} = \{y \in \mathbb{R}^n\colon |y-x_1| > 11\sqrt{n}2^{-\mu}\}$. Then

$$|\textstyle\int[K(x_1,y)-K(x_2,y)](a(y)-a(x_1))\overline{\varphi}_r(y-x_1)\,dy| \le$$

$$C\Big\{\int_{\mathscr{R}} \frac{|x_1-x_2|^\epsilon}{|y-x_1|^{n+\epsilon}}|a(y)-a(x_1)|\,dy + \int_{\mathscr{T}} \frac{|x_1-x_2|^\epsilon}{|y-x_1|^{n+\epsilon}}|Q|^{-1/2}\,dy\Big\} \le$$

$$C\Big\{|Q|^{-(1/2)-(1/n)}\int_{\mathscr{R}} \frac{|x_1-x_2|^\epsilon}{|y-x_1|^{n+\epsilon-1}}\,dy + |Q|^{-1/2}|x_1-x_2|^\epsilon 2^{\mu\epsilon}\Big\} \le$$

$$C|Q|^{-1/2}(2^\mu|x_1-x_2|)^\epsilon \le \tag{2.7}$$

$$C|Q|^{-1/2}(2^\mu|x_1-x_2|)^\epsilon\{(1+2^\mu|x_1|)^{-n-\epsilon} + (1+2^\mu|x_2|)^{-n-\epsilon}\},$$

provided $\epsilon < 1$. When $\epsilon = 1$ inequality (2.7) fails. It is for this reason that we have to modify the argument at this point and obtain a $(\beta,n+1)$ molecule instead of a $(1,n+1)$ molecule. We leave the details to the reader.

We estimate the second summand in (2.5) as follows: Let $\mathscr{U} = \{y\in\mathbb{R}^n\colon 0 < |y-x_1| \le 20|x_1-x_2|\}$. Then

$$|\textstyle\int K(x_1,y)(a(y)-a(x_1))\varphi_r(y-x_1)\,dy| \le C\displaystyle\int_{\mathscr{U}}|K(x_1,y)||a(y)-a(x_1)|\,dy \le$$

$$C\int_{\mathcal{U}} \frac{|y-x_1|}{|y-x_1|^n} |Q|^{-(1/2)-(1/n)} dy \le C|Q|^{-(1/2)-(1/n)}|x_1-x_2| \le$$

$$C|Q|^{-1/2}(2^{\mu}|x_1-x_2|)^{\epsilon}\{(1+2^{\mu}|x_1|)^{-n-\epsilon} + (1+2^{\mu}|x_2|)^{-n-\epsilon}\}.$$

The third summand in (2.5) is estimated in exactly the same way.

One of the remaining two cases is given by the inequalities $|x_1| \ge 10\sqrt{n}2^{-\mu}$ and $|x_2| < 10\sqrt{n}2^{-\mu}$; in the other the roles of x_1 and x_2 are reversed. Assume we have the first of this set of inequalities. Since $|x_1-x_2| < 2^{-\mu}$ we then have $|x_2| = |x_1+x_2-x_1| \ge |x_1|-|x_2-x_1| \ge 10\sqrt{n}2^{-\mu} - 2^{-\mu} \ge 9\sqrt{n}2^{-\mu}$. It is clear that a slight modification of the argument used in Case 1 gives us (2.4). This proves Theorem (1.15).

§3.**Proof of Theorem (1.16).** We will use a function $\theta \in \mathcal{D}$ that is radial satisfying (Supp θ) $\subset \{x \in \mathbb{R}^n: |x| \le 1\}$, $\int \theta(x)\, dx = 0$, and

$$\sum_{j=-\infty}^{\infty} \hat{\theta}(2^{-j}\xi) = 1$$

for $\xi \ne 0$. It is easy to see that such a function θ exists: Let $\Phi \in \mathcal{D}$ be radial, supported in $\{x \in \mathbb{R}^n: |x| \le 1/2\}$ and such that $\hat{\Phi}(0) = 1$. We then put $\theta(x) = \Phi(x) - 2^{-n}\Phi(x/2)$. Then $\hat{\theta}(\xi) = \hat{\Phi}(\xi) - \hat{\Phi}(2\xi)$ and, thus, $0 = \hat{\theta}(0) = \int \theta(x)\, dx$. Moreover, if $\xi \ne 0$,

$$\sum_{k=-N}^{N} \hat{\theta}(2^k\xi) = \sum_{k=-N}^{N} \{\hat{\Phi}(2^k\xi)-\hat{\Phi}(2^{k+1}\xi)\} = \hat{\Phi}(2^{-N}\xi)-\hat{\Phi}(2^{N+1}\xi)$$

and this last difference tends to $\hat{\Phi}(0) - 0 = 1$ as $N \to \infty$.

For $f \in \dot{F}_p^{\alpha,q}$ $(0 < \alpha < \beta,\ 1 < p,q < \infty)$ let $\Delta_j(f) \equiv f*\theta_j$, where $\theta_j(x) = 2^{jn}\theta(2^jx)$. Then $\sum_{-\ell}^{\ell} \Delta_j f$ converges to f in the $\dot{F}_p^{\alpha,q}$ norm as $\ell \to \infty$. Since $\tilde{T}$ is continuous on $\dot{F}_p^{\alpha,q}$, we have

$$\tilde{T}f = \sum_{-\infty}^{\infty} \tilde{T}\Delta_j f$$

(with convergence in the $\dot{F}_p^{\alpha,q}$ norm). Let $K_j(x,y)$ be the kernel of $T\Delta_j$; then $K_j(x,y) = T(\theta_j(\cdot - y))(x)$. Since $|Q|^{1/2}\theta_j(u - y)$ is an absolute constant times a smooth (1,0) atom (associated with a dyadic cube Q of side length 2^{-j} containing y), $K_j(x,y)$ is a $(\beta,n+\epsilon)$ molecule associated with the cube Q. Hence,

$$|K_j(x,y)| \le C|Q|^{-1/2}\frac{|Q|^{-1/2}}{(1+\ell(Q)^{-1}|x-y|)^{n+\epsilon}}$$

$$= C\,\frac{2^{jn}}{(1+2^j|x-y|)^{n+\epsilon}},$$

and

$$|K_j(x,y)-K_j(x',y)| \le C(2^j|x-x'|)^{\beta}\Big[\frac{2^{jn}}{(1+2^j|x-y|)^{n+\epsilon}}+\frac{2^{jn}}{(1+2^j|x'-y|)^{n+\epsilon}}\Big].$$

Thus,

$$|K(x,y)| \le \sum_{j=-\infty}^{\infty} |K_j(x,y)| \le C\sum_{j=-\infty}^{\infty}\frac{2^{jn}}{(1+2^j|x-y|)^{n+\epsilon}}$$

$$\le C\sum_{j=-\infty}^{\log_2 1/|x-y|} 2^{jn} + C\sum_{\log_2 1/|x-y|}^{\infty} 2^{-j\epsilon}|x-y|^{-n-\epsilon}$$

$$\le C2^{n\log_2 1/|x-y|} + C|x-y|^{-n-\epsilon}2^{-\epsilon\log_2 1/|x-y|} \le C|x-y|^{-n}.$$

For $|x-y| \ge 2|x-x'|$ we have

$$|K(x,y)-K(x',y)| \le \sum_{j=-\infty}^{\infty} |K_j(x,y)-K_j(x',y)| \le$$

$$C\sum_{j=-\infty}^{\infty}(2^j|x-x'|)^{\beta}\Big[\frac{2^{jn}}{(1+2^j|x-y|)^{n+\epsilon}}+\frac{2^{jn}}{(1+2^j|x'-y|)^{n+\epsilon}}\Big] \le$$

$$C|x-x'|^{\beta}\Big\{\sum_{j=-\infty}^{\log_2 1/|x-y|} 2^{j(n+\beta)} + \sum_{\log_2 1/|x-y|}^{\infty} 2^{(\beta-\epsilon)j}|x-y|^{-n-\epsilon}\Big\}$$

$$\le C|x-x'|^{\beta}\Big\{|x-y|^{-n-\beta}+|x-y|^{-n-\epsilon}|x-y|^{(\epsilon-\beta)}\Big\} \le C|x-x'|^{\beta}|x-y|^{-n-\beta}$$

(observe that we are using the assumption $\beta < \epsilon$).

This ends the proof of theorem (1.16).

Remarks: (i) It would appear that a more natural formulation of Theorem (1.16) would be to assume that T is continuous from all of $\mathscr{D}$ to $\mathscr{D}'$ and then conclude that, if T maps (1,0) atoms to $(\beta, n+\epsilon)$ molecules, then the kernel of T satisfies (1) and (2). As Rodolfo Torres pointed out, however, this is not true. Consider the operator T mapping g into the constant function with value $\int g$. T is clearly continuous from $\mathcal{D}$ to $\mathcal{D}'$ and maps (1,0) smooth atoms to 0 (by (1.8) with N = 0). But the Schwartz kernel of T is $K(x,y) \equiv 1$. The difficulty comes from the fact that $\tilde{T}|_{\mathcal{D}} \neq T$ in this case, since $\tilde{T}$ is the zero operator.

It is, perhaps, worth noting that, since $\mathcal{D}_0$ has a one dimensional complement in $\mathcal{D}$, the above example is essentially the only obstruction to the formulation given in the last paragraph. More precisely, suppose $T: \mathcal{D} \to \mathcal{D}'$ is continuous and maps (1,0) atoms to $(\beta,n+\epsilon)$ molecules. Let $\tilde{T}$ be the ($\dot{F}_p^{\alpha,q}$-continuous) extension of $T|_{\mathcal{D}_0}$ to $\mathcal{D} \subset \dot{F}_p^{\alpha,q}$. Fix any $\rho \in \mathcal{D}$ satisfying $\int \rho = 1$. For arbitrary $g \in \mathcal{D}$, let $\mu_g = \int g$. Then

$$(T - \tilde{T})g = (T - \tilde{T})(g - \mu_g\rho) + \mu_g(T - \tilde{T})\rho.$$

But $g - \mu_g\rho \in \mathcal{D}_0$ and $T - \tilde{T}$ agree on $\mathcal{D}_0$. Letting $(T - \tilde{T})\rho = \eta \in \mathcal{D}'$, we see that $(T - \tilde{T})g = \mu_g\eta$, where $\eta \in \mathcal{D}'$ is independent of g. This is the only possibility for the discrepancy between T and $\tilde{T}$. Observe that the Schwartz kernel of $T - \tilde{T}$ is $K(x,y) = \eta(y)$. In natural examples it is likely that auxiliary information about T (e.g. L^2 boundedness) will guarantee that $\eta = 0$ and, thus, $T = \tilde{T}$.

(ii)If $\alpha = 0$ and $1 \leq p,q < \infty$, the additional condition $\int m(x)\,dx = 0$ is needed to obtain the extension of theorem (1.13) to this case. The assumption $T^*1 = 0$ guarantees that $\int Ta = 0$ if a(x) is a (1,0) atom (this is an application of Fubini's theorem). Hence, using theorem (1.10), we obtain a proof, without using duality and interpolation, that if $T \in CZK \cap WPB$ and $T1 = 0 = T^*1$, then T is bounded on $\dot{F}_p^{0,q}$ for $1 \leq p,q < \infty$ (In particular, this includes the cases $L^p = \dot{F}_p^{0,2}$, $1 < p < \infty$, and $H^1 = \dot{F}_1^{0,2}$). The cases $\alpha \geq 1$ or $0 < p,q < 1$ will be considered elsewhere.

(iii) It is not really necessary to exclude the cases $p = \infty$ or $q = \infty$ in theorems (1.14) and (1.15). Our results also hold in these limit cases. In fact, the only point that requires some additional attention is the definition of $\dot{F}_\infty^{\alpha,q}$ (and $\dot{f}_\infty^{\alpha,q}$): we cannot formally put $p = \infty$ in the definition given in §1. However, with the correct definition (see [5]) our arguments carry over unchanged.

Bibliography

[1] Coifman, R., R., and Meyer, Y., **Au Delá des Opérateurs Pseudo–Différentiels**, Vol. 57, Astérisque (1982), pp. 1–199.

[2] Coifman, R., R. and Weiss, Guido , **Extensions of Hardy Spaces and Their Use in Analysis**, Bull. Amer. Math. Soc., Vol. 83 (1977), pp. 569–645.

[3] David, G. and Journé, J.–L., **A Boundedness Criterion for Generalized Calderón–Zygmund Operators**, Ann. of Math., Vol. 120 (1984), pp. 371–397.

[4] Frazier, M. and Jawerth, B., **Decomposition of Besov Spaces**, Ind. Univ. Math. J., Vol. 34 (1985), pp. 777–799.

[5] Frazier, M. and Jawerth, B., **A Discrete transform and Decompositions of Distribution Spaces**, Preprint.

[6] Hörmander, L., **The Analysis of Linear Partial Differential Operators, I. Distribution Theory**, Grunl. d. Math. Wiss., 256, Springer Verlag (1983).

[7] Lemarié, P. G., **Continuité sur les Espaces de Besov des Operateurs Definis par des integrales Singuliers**, Ann. Inst. Fourier, Grenoble, T 35, Fasc 4 (1985), pp. 175–187.

[8] Meyer, M., **Continuité Besov de Certains Opérateurs Integraux Singuliers**, These de 3e Cycle, Orsay 1985.

[9] Meyer, Y., **Les Nouveaux Opérateurs de Calderón–Zygmund**, Colloque en l'Honneur de L. Schwartz, I, Vol. 131, Astèrisque (1985), pp. 237–254.

[10] Peetre, J., **New Thoughts on Besov Spaces**, Duke Univ. Math. Series I, Dept. of Math., Duke University.

[11] Taibleson, M. and Weiss, Guido, **The Molecular Characterization of Certain Hardy Spaces**, 77, Astérisque (1980), pp. 67–149.

The University of New Mexico and Washington University, St. Louis

LES ONDELETTES

S. Jaffard et Y. Meyer
Université Paris-Dauphine (Ceremade)
Place de Lattre de Tassigny
75775 Paris Cedex 16

1. INTRODUCTION.

Nous nous proposons de faire le point sur les ondelettes. Nous dirons d'abord ce que sont des ondelettes, à quoi elles servent, quelles sont les diverses familles d'ondelettes dont on dispose aujourd'hui et enfin quelles sont les limitations inhérentes à l'usage de telle ou telle famille.

Les ondelettes sont de nouvelles "fonctions élémentaires" d'une ou plusieurs variables réelles, permettant de donner des représentations simples, efficaces et robustes des fonctions les plus générales sous la forme de séries d'ondelettes.

Les séries d'ondelettes se rattachent aux séries de Fourier mais présentent par rapport à ces dernières un triple avantage

(1) On s'affranchit de la contrainte de périodicité inhérente à l'utilisation des séries de Fourier et on peut désormais décomposer des fonctions tout à fait arbitraires (d'une ou plusieurs variables réelles) en séries d'ondelettes

(2) les fonctions $\cos kx$ ou $\sin kx$ qui interviennent dans les séries de Fourier ne sont pas bien localisées (par rapport à la variable x) alors que les ondelettes ψ_λ, $\lambda \in \Lambda$, sont très bien localisées, compte tenu de leur caractère oscillant et de leur régularité

(3) les séries d'ondelettes constituent une analyse de la fonction vis-à-vis d'une échelle dimensionnelle. Une série d'ondelettes est une série double dont les indices sont, en une dimension, j et k et ont la signification intuitive suivante : si $j = 0$, l'ondelette est une "loupe" que l'on déplace en les points $k \in \mathbb{Z}$ pour voir des détails de longueur 1 tandis que si $j = 1$, les ondelettes correspondantes $\psi_{j,k}$ permettront de voir les détails de longueur $1/2$ qui ont échappé à l'analyse précédente. Si $j = 2$, les ondelettes correspondantes $\psi_{j,k}$ jouent cette fois le rôle d'un microscope... Lorsqu'on examinera des détails de dimension 2^{-j}, il conviendra de déplacer le "microscope" en les points $k2^{-j}$, $k \in \mathbb{Z}$, ce qui est intuitivement évident.

Très grossièrement l'analyse par les ondelettes peut se décrire comme étant une "analyse de Fourier locale, à toutes les échelles".

Les ondelettes seront des "fonctions spéciales" ψ_λ, indexées par un ensemble Λ

d'indices λ et ayant des propriétés de régularité, de localisation et d'oscillation telles que la suite ψ_λ , $\lambda \in \Lambda$, forme une base orthonormée de l'espace de référence $L^2(\mathbb{R}^n)$ mais permette également de décomposer les distributions et les fonctions de test qui leur sont associées.

Cela signifie que, grâce aux qualités des ψ_λ , $\lambda \in \Lambda$, l'identité fondamentale

$$f(x) = \sum_{\lambda \in \Lambda} c_\lambda \psi_\lambda(x) \tag{1.1}$$

où

$$c(\lambda) = \int_{\mathbb{R}^n} f(x)\bar{\psi}_\lambda(x)\,dx \tag{1.2}$$

doit pouvoir fonctionner dans à peu près tous les cadres fonctionnels usuels.

Les cadres fonctionnels que l'on utilise aujourd'hui permettent de préciser la forme bilinéaire (S,φ) qui décrit la valeur d'une distribution S sur une fonction de test φ . Voici un exemple. On suppose $1 < p < \infty$ et l'on appelle $W^{m,p}$ l'espace des fonctions de φ qui, ainsi que toutes leurs dérivées partielles $\partial^\alpha\varphi$ d'ordre $|\alpha| = \alpha_1 + \dots + \alpha_n \leqslant m$, appartiennent à l'espace de Lebesgue $L^p(\mathbb{R}^n;dx)$. Alors le dual de $W^{m,p}$ est $W^{-m,q}$ où $1/q + 1/p = 1$ et où cette fois les éléments $S \in W^{-m,q}$ sont des (sommes finies de) dérivées $\partial^\beta\varphi$ où $|\beta| \leqslant m$ et $\varphi \in L^q$.

La dualité entre ces deux espaces s'exprime par une inégalité

$$|(S,\varphi)| \leqslant C\,\|S\|_{W^{-m,q}}\,\|\varphi\|_{W^{m,p}}$$

où C est une constante (dépendant de choix numériques sur les normes utilisées). Ce que l'on attend des ondelettes est de pouvoir écrire les distributions $S \in W^{-m,q}$ grâce à l'algorithme (1.1). Les coefficients se calculent grâce à (1.2) et cela oblige les ondelettes à appartenir à $W^{m,p}$.

L'ensemble Λ que l'on utilise dans la construction des ondelettes est la réunion des réseaux $\Lambda_j = 2^{-j}\mathbb{Z}^n$ de $\mathbb{R}^n$ (on excepte 0 qui n'appartient pas à Λ). Cela signifie que Λ n'est pas un ensemble ordonné. L'approximation d'une fonction $f(x)$ définie sur $\mathbb{R}^n$ par des combinaisons linéaires d'ondelettes ψ_λ reflète et imite l'approximation de $\mathbb{R}^n$ par les Λ_j . C'est bien ce que signifie l'analyse dimensionnelle décrite par (3). C'est également ce que les physiciens travaillant en théorie quantique des champs exigeaient de ce qu'ils ont appelé "lattice approximation". Tout cela implique que les séries d'ondelettes des fonctions ou distributions des différents cadres fonctionnels B que l'on utilise doivent être des séries commutativement convergentes (que l'on appelle aussi des familles sommables ou des séries inconditionnellement convergentes) pour la norme de B .

Cela nous force à "prendre nos désirs pour des réalités" : caractériser l'appartenance d'une fonction ou d'une distribution f à l'un des cadres fonctionnels (ou espaces de Banach) B d'utilisation courante au seul vu de la suite des modules des coefficients d'ondelettes correspondants.

Nous verrons que les ondelettes peuvent relever ce défi, à condition d'exclure certains espaces fonctionnels que l'on sait, pour des raisons abstraites, ne pouvoir posséder de base inconditionnelle. Les espaces exclus sont $L^1(\mathbb{R}^n)$, son dual $L^\infty(\mathbb{R}^n)$ et ensuite les espaces $C^1(\mathbb{R}^n)$ (fonctions continûment dérivables)... En revanche sont admis les espaces de Hölder homogènes $C^r(\mathbb{R}^n)$, $r \notin \mathbb{N}$. Rappelons que $f \in C^r(\mathbb{R}^n)$ signifie l'existence, si $0 < r < 1$, d'une constante C telle que $|f(x+h)-f(x)| \leqslant C|h|^r$ pour tout $x \in \mathbb{R}^n$ et tout $h \in \mathbb{R}^n$. Si $1 < r < 2$, on remplace cette condition par $|f(x+h)+f(x-h)-2f(x)| < C|h|^r$ etc... (Cette dernière condition, écrite pour $r = 1$, définit la classe de Zygmund).

Un second défi imposé à la famille ψ_λ, $\lambda \in \Lambda$, est de fournir, en un certain sens, une analyse de Fourier locale permettant de détecter la régularité ou l'irrégularité locale (ou même ponctuelle) d'une fonction ou d'une distribution.

Un troisième défi est la simplicité de la construction des ψ_λ, $\lambda \in \Lambda$. En dimension un, on cherchera des bases de la forme $2^{j/2}\psi(2^j x-k)$ où $j \in \mathbb{Z}$, $k \in \mathbb{Z}$. Cet algorithme reflète, encore une fois, l'analyse dimensionnelle décrite plus haut. La fonction $\psi(x)$ sera appelée l'<u>ondelette analysante</u>. Elle doit être très régulière, très bien localisée et très oscillante pour que les fonctions $\psi_{j,k}(x) = 2^{j/2}\psi(2^j x-k)$, $j \in \mathbb{Z}$, constituent une base orthonormée de $L^2(\mathbb{R})$ et également une base inconditionnelle des autres espaces fonctionnels d'usage courant (espaces $W^{m,p}$, espaces C^r, $r \notin \mathbb{N}$, etc.).

Les conditions de régularité sont que, pour un certain entier m, $\psi(x)$ soit de classe C^m. Les conditions de localisation sont que $\psi(x)$ ainsi que toutes ses dérivées jusqu'à l'ordre m soient à décroissance rapide à l'infini. (Si $m = +\infty$, $\psi(x)$ appartiendra à la classe $S(\mathbb{R}^n)$ de Schwartz). Enfin les conditions d'oscillation sont $\int_{-\infty}^{\infty} x^k\psi(x)\,dx = 0$ si $0 \leqslant k \leqslant m$.

Tout cela s'étend également à des ondelettes analysantes en n variables réelles. Elles seront alors au nombre de $q = 2^n-1$, seront notées $\psi_1,\dots,\psi_q$ et les autres ondelettes seront alors données par la formule $2^{nj/2}\psi_\ell(2^j x-k)$ où $j \in \mathbb{Z}$, $k \in \mathbb{Z}^n$ et $1 \leqslant \ell \leqslant q$. Cette collection sera la base orthonormée recherchée.

Nous <u>ne savons pas</u> à l'heure actuelle, construire <u>toutes</u> les bases orthonormées relevant des défis précédents. En revanche nous connaissons un certain nombre d'exemples. Dans tous ces exemples, les ondelettes proviennent d'une <u>analyse graduée</u>. Cette notion, présentée au troisième paragraphe, généralise les suites emboîtées de fonctions splines associées à l'approximation de $\mathbb{R}^n$ par les réseaux emboîtés Λ_j, $j \geqslant 0$. Précisément les noeuds des splines $f \in V_j$ seront les points $\lambda \in \Lambda_j$, dans l'une des constructions que P.G. Lemarié a obtenue. En général la relation est plus abstraite.

Le passage d'une analyse graduée à la base d'ondelettes correspondant est une opération non-linéaire, explicitée en dimension un par l'auteur de ces liges et découverte par K. Gröchenig (Vienne) en dimension supérieure. Les lemmes d'orthogonalisation du second paragraphe y jouent un rôle fondamental.

2. BASES ORTHONORMEES ET BASES INCONDITIONNELLES.

Dans cette section H désignera un espace de Hilbert séparable. Cela signifie que H contient une partie dénombrable dense ou encore que, soit H est de dimension finie, soit isomorphe à $\ell^2(\mathbb{N})$.

Nous désignerons par $(x|y)$ le produit scalaire entre $x \in H$ et $y \in H$. Dans la définition qui suit, H est soit un espace de Hilbert, soit, plus généralement, un espace de Banach.

DEFINITION 1. Une suite $e_o, e_1, \ldots, e_k, \ldots$ de vecteurs d'un espace de Banach H est une base inconditionnelle de H si tout vecteur $x \in H$ s'écrit $x = \alpha_o e_o + \alpha_1 e_1 + \ldots + \alpha_k e_k + \ldots$ où les coefficients $\alpha_o, \alpha_1, \ldots$ sont tels que la série écrite soit une famille sommable, convergeant vers x pour la norme de H et si ces coefficients $\alpha_o, \alpha_1, \ldots, \alpha_k, \ldots$ sont alors uniques.

Si H est un espace de Banach et si $e_o, \ldots, e_k, \ldots$ est une base inconditionnelle de H , alors les suites des coordonnées α_k , $k \in \mathbb{N}$, des vecteurs de H sont caractérisées par une condition ne portant que sur les modules $|\alpha_k|$, $k \in \mathbb{N}$ (et, en outre, cette condition est "croissante" au sens que, si elle est satisfaite pour la suite $|\beta_k|$, elle l'est automatiquement pour toute suite α_k , $k \in \mathbb{N}$, telle que $|\alpha_k| \leq |\beta_k|$) . Si H est un espace de Hilbert et si e_k , $k \in \mathbb{N}$ est une base inconditionnelle de H , la condition en question $\sum_o^\infty |\alpha_k|^2 \|e_k\|^2 < \infty$. On renormalise alors les e_k en les remplaçant par $e_k/\|e_k\|$. La donnée d'une base inconditionnelle ainsi renormalisée revient à celle d'un isomorphisme (non isométrique, en général) $T : \ell^2(\mathbb{N}) \to H$. Si T est isométrique, alors e_k , $k \in \mathbb{N}$, est une base hilbertienne.

Nous nous proposons ensuite de transformer la base $(e_o, e_1, \ldots, e_k, \ldots)$ en une base orthonormée $(f_o, f_1, \ldots, f_k, \ldots)$ mais en respectant la symétrie inhérente à la définition d'une base inconditionnelle. On demande donc que l'algorithme utilisé transforme $e_{\pi(0)}, \ldots, e_{\pi(k)}, \ldots$ en $f_{\pi(0)}, \ldots, f_{\pi(k)}$ quelle que soit la permutation π utilisée. Cela signifie que l'on s'interdit d'utiliser l'algorithme de Gram-Schmidt car il ne respecte pas les symétries. Voici un second algorithme qui remplit les conditions exigées.

On considère l'opérateur $T : H \to H$ défini par $T(x) = \sum_o^\infty \langle x|e_k\rangle e_k$. La convergence de cette série résulte de l'équivalence entre $\|x\|$ et $\left(\sum_o^\infty |\langle x|e_k\rangle|^2\right)^{1/2}$.

Cette dernière équivalence est une conséquence simple de la définition 1.

Si l'opérateur T était l'identité, les e_k constitueraient une base orthonormée de H. On a, en fait,

$$c_1 1 \leqslant T \leqslant c_2 1$$

pour deux constantes c_2 et c_1 strictement positives. Cela entraîne que $T^{-1/2}$ a un sens (calcul symbolique sur les opérateurs auto-adjoints positifs) et l'on a, finalement, la base hilbertienne cherchée $f_o, f_1, \ldots, f_k, \ldots$ sous la forme

$$f_k = T^{-1/2}(e_k)$$

On désigne par $G = ((\langle e_j\, e_k\rangle))_{\mathbb{N}\times\mathbb{N}}$ la matrice de Gram des vecteurs e_j, $j \in \mathbb{N}$. Cette matrice G est la matrice de l'opérateur T, dans la base des e_k, $k \in \mathbb{N}$. On fait évidemment opérer G sur $\ell^2(\mathbb{N})$ et l'on a, au sens des matrices définies-positives $\gamma_1 1 \leqslant G \leqslant \gamma_2 1$, où $\gamma_2 \geqslant \gamma_1 > 0$. Cela permet de définir $G^{-1/2}$. Cette matrice $G^{-1/2}$ est la matrice de $T^{-1/2}$ dans la base des e_k $(k \in \mathbb{N})$. On a finalement $f_j = \sum_o^\infty \gamma(j,k) e_k$ où les $\gamma(j,k)$ sont les coefficients de $G^{-1/2}$.

Dans les applications, nous aurons besoin du lemme suivant qui est un perfectionnement de résultats classiques en analyse numérique.

LEMME 1. Soit T un ensemble muni d'une distance $d(s,t)$ vérifiant la propriété suivante

il existe un entier $n \geqslant 1$ et une constante C tels que, pour tout $R \geqslant 1$, le nombre des points $t \in T$ vérifiant $d(s,t) \leqslant R$ ne dépasse pas $C R^n$. (2.1)

Soit A une matrice dont les coefficients $\alpha(s,t)$, $s \in T$, $t \in T$, vérifient, pour tout entier $N \geqslant 1$

$$|\alpha(s,t)| \leqslant C_N(1 + d(s,t))^{-N} \tag{2.2}$$

Supposons en outre que, pour deux constantes $c_2 \leqslant c_1 > 0$,

$$c_1 1 \leqslant A \leqslant c_2 1 \tag{2.3}$$

(au sens des matrices définies-positives).

Alors les coefficients $\gamma(s,t)$, $s \in T$, $t \in T$, de la matrice $A^{-1/2}$ vérifient

$$|\gamma(s,t)| \leqslant C'_N(1 + d(s,t))^{-N} \tag{2.4}$$

pour tout entier $N \geqslant 1$.

Nous appliquerons ce lemme à la matrice de Gram associée à une base inconditionnelle e_s , $s \in T$, d'un espace de Hilbert H . Alors (2.2) signifie que les e_s sont presque-orthogonaux et la conclusion est que les vecteurs orthogonaux $f_s = \sum \gamma(s,t)e_t$ utilisent principalement les $t \in T$ proches de s .

3. ESPACES EMBOITES DE SPLINES GENERALISES.

L'espace fonctionnel $L^2(\mathbb{R}^n;dx)$ est l'espace de référence dans les définitions qui suivent. Son caractère d'espace "pivot" vient de ce que la dualité entre distributions et fonctions de test s'écrit en utilisant le produit scalaire interne de l'espace $L^2(\mathbb{R}^n;dx)$ - du moins si la fonction de test est à valeurs réelles.

On a, en effet, si $\varphi(x) \in \mathcal{D}(\mathbb{R}^n)$ et $S \in \mathcal{D}'(\mathbb{R}^n)$

$$(S,\varphi) = \int_{\mathbb{R}^n} S(x)\,\varphi(x)\,dx$$

et

$$\langle u,v\rangle = \int_{\mathbb{R}^n} u(x)\,\bar{v}(x)\,dx$$

si u et v appartiennent à $L^2(\mathbb{R}^n)$.

Nous commencerons donc par présenter la définition d'une <u>analyse graduée</u> de $L^2(\mathbb{R}^n)$. Nous verrons ensuite que, sous certaines conditions très simples, cette analyse graduée permet d'analyser les espaces fonctionnels les plus variés.

DEFINITION 2. <u>Une analyse graduée de</u> $L^2(\mathbb{R}^n)$ <u>est, par définition, une suite croissante</u> V_j , $j \in \mathbb{Z}$, <u>de sous-espaces fermés de</u> $L^2(\mathbb{R}^n)$, <u>possédant, en outre, les quatre propriétés suivantes</u> :

$\bigcap_{-\infty}^{\infty} V_j = \{0\}$ <u>et</u> $\bigcup_{-\infty}^{\infty} V_j$ <u>est dense dans</u> $L^2(\mathbb{R}^n)$ (3.1)

$f(x) \in V_j \Leftrightarrow f(2x) \in V_{j+1}$, <u>pour n'importe quelle fonction</u> f <u>de</u> V_j (3.2)

<u>pour toute fonction</u> $f(x)$ <u>appartenant à</u> V_o <u>et tout</u> $k \in \mathbb{Z}^n$, $f(x-k)$ <u>appartenant à</u> V_o (3.3)

<u>il existe (au moins) une fonction</u> $g(x) \in V_o$ <u>telle que la suite</u> $g(x-k)$, $k \in \mathbb{Z}^n$, <u>soit une base inconditionnelle de</u> V_o . (3.4)

Voici des exemples. Nous commencerons par des analyses graduées où la définition des V_j est de type géométrique. Ensuite nous envisagerons des descriptions plus subtiles utilisant la transformation de Fourier.

<u>Exemple 1.</u> L'espace V_o est le sous-espace de $L^2(\mathbb{R})$ formé des fonctions en escalier $f(x)$ telles que $f(x) = a_k$ si $k \leq x < k+1$, $k \in \mathbb{Z}$. Alors $g(x)$ est la fonction caractéristique de $[0,1[$ et la base des $g(x-k)$, $k \in \mathbb{Z}$, est orthonormée.

<u>Exemple 2.</u> Cette fois V_o est le sous-espace de $L^2(\mathbb{R})$ composé des fonctions

continues et dont la restriction à chaque intervalle $[k,k+1]$, $k \in \mathbb{Z}$, est une fonction affine $a_k x + b_k$. Ici $g(x)$ est, de façon naturelle, la "fonction triangle" $\Delta(x) = \sup(0, 1-|x|)$.

Exemple 3. On part d'un entier $m \geqslant 1$ et $V_o = V_o^{(m)}$ est le sous-espace de $L^2(\mathbb{R})$ composé des fonctions de classe C^{m-1} dont la restriction à chaque intervalle $[k,k+1]$, $k \in \mathbb{Z}$, est un polynôme $P_k(x)$ de degré $\leqslant m$. Alors un choix de $g(x)$ est la fonction qui est appelée le "basic spline" et qui est le produit de convolution $\chi \star \chi \star \ldots \star \chi$ (m+1 termes), χ est la fonction caractéristique de $[0,1]$.

Il faut observer que si $g(x)$ convient, il en sera de même pour $g(x-\ell)$ pour tout $\ell \in \mathbb{Z}$. Appelons $g_m(x)$ le "basic spline" de l'exemple 3. Alors la fonction triangle $\Delta(x)$ de l'exemple 2 n'est autre que $g_2(x+1)$.

On vérifie que $g_m(x) \geqslant 0$, le support de $g_m(x)$ est l'intervalle $[0,m+1]$ et que $\sum_{-\infty}^{\infty} g_m(x-k) = 1$.

Exemple 4. Nous désignerons, cette fois, par $V_o \subset L^2(\mathbb{R})$ le sous-espace fermé composé des fonctions $f(x)$ dont la transformée de Fourier $\hat{f}(\xi) = \int_{-\infty}^{\infty} e^{ix\xi} f(x)dx$ est supportée par l'intervalle $[-\pi,\pi]$. Alors, en posant $g(x) = \dfrac{\sin \pi x}{\pi x}$, la suite $g(x-k)$, $k \in \mathbb{Z}$, est une base orthonormée de V_o . La fonction $g(x)$ s'appelle le "sinus cardinal".

Exemple 5. Cet exemple est une correction du précédent où V_o est modifié de sorte que la fonction $g(x)$ appartienne à la classe $S(\mathbb{R})$ de Schwartz.

On part d'une fonction $\theta(\xi)$ qui est indéfiniment dérivable sur toute la droite réelle, paire, égale à 1 sur $[0, \frac{2\pi}{3}]$, à 0 si $\xi \geqslant \frac{4\pi}{3}$, qui est comprise entre 0 et 1 sur l'intervalle $[\frac{2\pi}{3}, \frac{4\pi}{3}]$ et qui y vérifie $\theta^2(\xi) + \theta^2(2\pi-\xi) = 1$.

On appelle φ la fonction définie par $\varphi(x) = \frac{1}{2\pi} \int_{-\infty}^{\infty} e^{ix\xi}\theta(\xi)d\xi$. Ainsi θ est la transformée de Fourier de φ . Les fonctions $\varphi(x-k)$, $k \in \mathbb{Z}$, forment une suite orthonormée dans $L^2(\mathbb{R})$ et on désigne par V_o le sous-espace fermé de $L^2(\mathbb{R})$ engendré par les $\varphi(x-k)$, $k \in \mathbb{Z}$. Alors (3.2) permet de construire une analyse graduée à partir de V_o .

Exemple 6. Nous donnons maintenant des exemples dans $L^2(\mathbb{R}^2)$. Il y a un moyen systématique pour prduire une analyse graduée $\tilde{V}_j$, $j \in \mathbb{Z}$, de $L^2(\mathbb{R}^2)$ à partir d'une analyse graduée V_j , $j \in \mathbb{Z}$, de $L^2(\mathbb{R})$. On forme le produit tensoriel (algébrique) $V_j \otimes V_j$ que l'on complète en le munissant de la norme induite par celle de $L^2(\mathbb{R})$ et l'on obtient $\tilde{V}_j$. Une base inconditionnelle de $\tilde{V}_o$ est alors formée des fonctions $g(x-k)g(y-\ell)$, $k \in \mathbb{Z}$, $\ell \in \mathbb{Z}$.

Exemple 7. On part du réseau hexagonal Γ engendré par les vecteurs $(1\cdot 0)$ et

$[\frac{1}{2}, \frac{\sqrt{3}}{2}]$. Ce réseau Γ remplacera le réseau $\mathbb{Z}^2$ de la définition 2. Les points Γ sont naturellement les sommets de triangles équilatéraux $T \in \mathcal{T}$ réalisant un pavage du plan. On désigne alors par V_0 le sous-espace fermé de $L^2(\mathbb{R}^2)$ composé des fonctions continues sur $\mathbb{R}^2$ dont la restriction à chaque triangle $T \in \mathcal{T}$ est une fonction affine.

Cette liste d'exemples n'est pas exhaustive.

DEFINITION 3. Soit $r \geq 0$ un entier. Nous dirons qu'une analyse graduée V_j, $j \in \mathbb{Z}$, de $L^2(\mathbb{R}^n)$ est r-régulière si l'on peut choisir la fonction g qui intervient dans la définition 2 de sorte que

$g(x)$ soit une fonction de classe C^r (c'est-à-dire que toutes les dérivées partielles $\partial^\alpha g(x)$ où $\alpha = (\alpha_1,\ldots,\alpha_n)$ et $|\alpha| = \alpha_1 +\ldots+ \alpha_n \leq r$ soient continues) et que (3.5)

ces dérivées partielles $\partial^\alpha g(x)$ aient une décroissance rapide à l'infini. (3.6)

En d'autres termes, on demande que $g(x)$ ait une bonne régularité et soit bien localisée autour de 0 . Les analyses graduées des exemples 1 et 4 ne sont pas r-régulières.

Désignons par $E_j : L^2(\mathbb{R}^n) \to V_j$ l'opérateur de projection orthogonale sur V_j. Dès que l'analyse graduée est r-régulière, les opérateurs E_j se prolongent à tous les espaces fonctionnels classiques, à condition que la régularité ou l'irrégularité que ces espaces permettent de décrire ne dépasse par R. De plus, si f appartient à l'un de ces espaces de Banach classiques B, alors $E_j(f)$ converge vers f quand j tend vers l'infini.

4. CONSTRUCTION DE BASES D'ONDELETTES.

Nous souhaitons construire une base d'ondelette ψ_λ, $\lambda \in \Lambda$, à partir d'une analyse graduée donnée par la suite croissante V_j, $j \in \mathbb{Z}$. L'algorithme conduisant aux ondelettes analysantes $\psi_1,\ldots,\psi_q$, $q = 2^n-1$, devra permettre de conserver les mêmes propriétés de régularité et de localisation dont on dispose sur la fonction $g(x)$ intervenant dans la définition de l'analyse graduée.

Voici les grandes lignes de la construction.

On appelle $W_j \subset V_{j+1}$ le complément orthogonal de V_j dans V_{j+1} et l'on peut alors écrire

$$L^2(\mathbb{R}^n) = \bigoplus_{-\infty}^{\infty} W_j \qquad (4.1)$$

On a l'équivalence

$$f(x) \in W_o \Leftrightarrow f(2^j x) \in W_j \quad . \tag{4.2}$$

Le théorème fondamental qui suit a été établi par K. Gröchenig (Vienne) en mars 1987.

THEOREME 1. _Soit_ V_j , $j \in \mathbb{Z}$, _une analyse graduée_ r _-régulière. Il existe alors_ $q = 2^n - 1$ _fonctions_ $\psi_1,\dots,\psi_q$, _de classe_ C^r _telles que la collection complète_

$$2^{nj/2}\psi_1(2^j x-k) \ ,\dots,\ 2^{nj/2}\psi_q(2^j x-k) \quad , \ j \in \mathbb{Z} \ , \ k \in \mathbb{Z}^n \ , \tag{4.3}$$

soit une base orthonormée de $L^2(\mathbb{R}^n)$.

De plus toutes les dérivées $\partial^\alpha\psi_1(x),\dots,\partial^\alpha\psi_q(x)$ _où les_ $|\alpha| \leq r$ _ont une décroissance rapide à l'infini._

Essayons de décrire la preuve de ce résultat.

Il suffit, en fait, que $\psi_1(x-k),\dots,\psi_q(x-k)$, $k \in \mathbb{Z}^n$, constitue une base orthonormée de W_o .

On part de la base inconditionnelle $g(x-k)$, $k \in \mathbb{Z}^n$, de V_o . On lui applique le lemme 1 et l'on obtient ainsi une base orthonormée $\varphi(x-k)$, $k \in \mathbb{Z}^n$, de V_o . On dispose également d'une base orthonormée $2^{n/2}\varphi(2x-k)$, $k \in \mathbb{Z}^n$, de V_1 . En posant $\varphi_2(x) = 2^{n/2}\varphi(2x)$, cette nouvelle base s'écrit $\varphi_2(x-k/2)$ et ces fonctions sont centrées aux points du réseau $2^{-1}\mathbb{Z}^n$.

Il est naturel que la base recherchée du complément orthogonal W_o de V_o dans V_1 soit composée de fonctions centrées en $2^{-1}\mathbb{Z}^n \setminus \mathbb{Z}^n$ (comme ce serait le cas, en dimension finie, en appliquant le théorème de la base incomplète). Or $2^{-1}\mathbb{Z}^n \setminus \mathbb{Z}^n$ se compose des 2^n-1 classes $r + \mathbb{Z}^n$ où $r = \left(\frac{\varepsilon_1}{2},\dots,\frac{\varepsilon_n}{2}\right)$, $\varepsilon_1 = 0$ ou $1,\dots,$ $\varepsilon_n = 0$ ou 1 et l'on exclut $\varepsilon_1 = \dots = \varepsilon_n = 0$. Cela explique intuitivement pourquoi la base orthonormée de W_o est de la forme $\psi_1(x-k),\dots,\psi_q(x-k)$, $q = 2^n-1$, $k \in \mathbb{Z}^n$. Mais cela ne permet pas de calculer les fonctions $\psi_1,\dots,\psi_q$.

Si $n = 1$, on procède comme suit. Il existe une fonction $m_o(\xi)$, indéfiniment dérivable sur $\mathbb{R}$, 2π -périodique et telle que $\hat{\varphi}(2\xi) = m_o(\xi)\hat{\varphi}(\xi)$.

Alors ψ sera définie par $\hat{\psi}(2\xi) = m_1(\xi)\hat{\varphi}(\xi)$ où $m_1(\xi) = e^{i\xi}\,\overline{m_o(\xi+\pi)}$.

Une des plus belles applications du théorème 1 est le théorème suivant, dû à I. Daubechies.

THEOREME 2. _Pour tout entier_ $m \in \mathbb{N}$, _il existe deux fonctions de classe_ C^m , _à support compact,_ φ _et_ ψ , _telles que la collection_

$$\varphi(x-k) \quad \text{et} \quad 2^{j/2}\psi(2^j x-k) \quad , \ j \in \mathbb{N} \ , \ k \in \mathbb{Z}$$

soit une base orthonormée de $L^2(\mathbb{R},dx)$.

Illustrons le théorème 2. On part d'une distribution S d'ordre $r < m$, appartenant à $\mathcal{D}'(\mathbb{R})$. Cela signifie que S est une forme linéaire continue sur l'espace des fonctions de classe C^r à support compact. Nous supposerons pour simplifier que r n'est pas entier. (Alors C^r est défini à l'aide de l'échelle des espaces de Hölder).

On commence par former <u>une image floue</u> S_o de S, à savoir $S_o(x) = \sum_{-\infty}^{\infty} c_k\varphi(x-k)$ où $c_k = \langle S,\varphi_k\rangle$ et $\varphi_k(x) = \varphi(x-k)$. Les coefficients c_k donnent une première lecture de S aux points entiers $k \in \mathbb{Z}$.

On veut naturellement en savoir plus et l'on passe aux points demi-entiers. La meilleure approximation de $S(x)$ que l'on obtient est notée $S_1(x)$. Elle est obtenue à partir de $S_o(x)$ en ajoutant de petites fluctuations bien localisées. Ces petites fluctuations sont précisément les ondelettes et permettent de mieux rendre compte de la structure intime ou complexe de la distribution S.

Plus précisément, on aura

$$S(x) = S_o(x) + (S_1(x)-S_o(x)) + (S_2(x)-S_1(x)) + \ldots + (S_{j+1}(x)-S_j(x)) + \ldots \qquad (4.4)$$

où $S_{j+1}(x)-S_j(x) = \sum_{-\infty}^{\infty} b(k,j)2^{j/2}\psi(2^jx-k)$, $b(k,j) = \langle S,\psi_{j,k}\rangle$ et $\psi_{j,k}(x) = 2^{j/2}\psi(2^jx-k)$.

Mais il y a mieux. Les <u>coefficients d'ondelettes</u> $b(k,j)$ de notre distribution $S(x)$ d'ordre r sont caractérisés par la condition que, pour tout $R \geqslant 1$,

$$\sum_{o}^{\infty} \sum_{-R2^j}^{R2^j} |b(k,j)|2^{j(\frac{1}{2}-r)} < +\infty \quad . \qquad (4.5)$$

De plus la dualité entre une distribution S d'ordre $r > 0$ et une fonction de test u de classe C^r, à support compact est donnée par

$$(S,\bar{u}) = \sum_{-\infty}^{\infty} c_k\bar{\gamma}_k + \sum_{o}^{\infty}\sum_{-\infty}^{\infty} b(k,j)\bar{\beta}(k,j) \qquad (4.6)$$

lorsque $c_k = (S,\bar{\varphi}_k)$, $b(k,j) = (S,\bar{\psi}_{j,k})$ et $\gamma_k = (u,\bar{\varphi}_k)$, $\beta(k,j) = (u,\bar{\psi}_{j,k})$.

Enfin la série (4.6) est absolument convergente.

Le théorème de Daubechies s'étend immédiatement en plusieurs dimensions. Il en est donc de même pour l'analyse, en séries d'ondelettes, des distributions S d'ordre r (en supposant toujours r non entier).

Par exemple la base de Daubechies bi-dimensionnelle se compose des fonctions $\varphi(x-k)\varphi(y-\ell)$, $k \in \mathbb{Z}$, $\ell \in \mathbb{Z}$, des fonctions $2^j\varphi(2^jx-k)\psi(2^jy-\ell)$, $2^j\psi(2^jx-k)\varphi(2^jy-\ell)$ et des fonctions $2^j\psi(2^jx-k)\psi(2^jy-\ell)$, $j \in \mathbb{N}$, $k \in \mathbb{Z}$, $\ell \in \mathbb{Z}$.

5. CONCLUSION : LES RECHERCHES EN COURS.

Nous appartenons à une équipe pluridisciplinaire. Notre chef d'orchestre est A. Grossmann et il déborde d'optimisme. Notre équipe comprend des scientifiques dont les formations et les goûts sont très variés (prospection pétrolière, musique, analyse de la parole, mécanique quantique,...). Nous ne pouvons décrire ici tous nos projets scientifiques mais nous nous limiterons à quelques recherches purement mathématiques conduites par les auteurs et par P.G. Lemarié (Ceremade).

Il s'agit de construire des ondelettes ψ_λ adaptées à la dualité $(\mathcal{D}(\Omega),\mathcal{D}'(\Omega))$ entre fonctions de test à support compact contenu dans un ouvert borné $\Omega \subset \mathbb{R}^n$ et distributions. Là encore on se limite à des fonctions u appartenant à $C^r(\Omega)$, $r > 0$, $r \notin \mathbb{N}$ et aux distributions S d'ordre r. Alors par deux approches tout à fait différentes, Jaffard et Lemarié construisent deux bases (très différentes) ψ_λ, $\lambda \in \Lambda$, de $L^2(\Omega)$ qui permettent d'écrire de façon simple, robuste et efficace cette dualité.

Ces travaux sont en cours de rédaction.

Références.

[1] I. Daubechies. Orthonormal bases of compactly supported wavelets. (To appear) AT & T BELL LABORATORIES, 600 Mountain Avenue, Murray Hill, NJ 07974.

[2] K. Gröchenig. Analyse multi-échelles et bases d'ondelettes. C.R. Acad. Sci. Paris, t. 305, Série 1, p. 13-15, 1987.

[3] S. Jaffard et Y. Meyer. Bases d'ondelettes dans des ouverts de $\mathbb{R}^n$. Soumis au Journal de Mathématiques Pures et Appliquées.

[4] S. Jaffard, Y. Meyer et O. Rioul. L'analyse par ondelettes. Pour la science. Septembre 1987. p. 28-37.

[5] P.G. Lemarié et Y. Meyer. Ondelettes et bases hilbertiennes. Revista Matematica Iberoamericana, 2, 1, (1986). Volume dédié à A. Calderon.

[6] S. Mallat. Multiresolution approximation and wavelets. GRASP lab. Dept. of Computer and Information Science, Univ. of Pennsylvania, Philadelphia PA 19104-6389 USA.

ON FUNCTIONS WITH DERIVATIVES IN H^1.

Svante Janson
Department of Mathematics
Uppsala University
Thunbergsvägen 3
S-752 38 UPPSALA, Sweden

1. Introduction and results.

Let IH^1 be the set of all locally integrable functions g on $\mathbb{R}^n$ such that $\nabla g \in H^1(\mathbb{R}^n)$ (i.e. the weak partial derivatives $D_j g = \partial g/\partial x_j \in H^1$, $j=1...n$).

When $n=1$, IH^1 is the set of primitive functions of functions in H^1;

$$IH^1 = \{ \int_{-\infty}^{x} f(t)dt + c : f \in H^1\} , \quad n=1 . \tag{1.1}$$

In general we have the related characterization

$$IH^1 = \{I f + c : f \in H^1\} , \quad n \geq 1 , \tag{1.2}$$

where I is fractional integration of order 1. (This follows because the Riesz transforms $R_j = D_j I$ are bounded on H^1, and $g = \Sigma I R_j D_j g$ cf. [5].) Note that all functions in IH^1 are bounded and continuous when $n=1$, but not when $n \geq 2$.

Remark. IH^1 equals the homogeneous Triebel-Lizorkin space $\dot{F}_1^{12}$. (See e.g. [6] for definitions.) We will not consider Triebel-Lizorkin spaces with other parameters here.

The main purpose of this paper is to characterize the functions φ that operate on IH^1 in the sense that $\varphi \circ g \in IH^1$ for every $g \in IH^1$. (We may consider either $\varphi : \mathbb{R} \to \mathbb{R}$ and only real functions in IH^1, or $\varphi : \mathbb{C} \to \mathbb{C}$ and complex functions in IH^1; there is no difference between these cases.)
Before stating the result we recall the following definition.

A function φ is Lipschitz if there exists a constant C such that

$$|\varphi(s) - \varphi(t)| \leq C|s - t| \quad \text{for all } s,t . \tag{1.3}$$

φ is locally Lipschitz if for every R there exists a C (depending on R) such that (1.3) holds when $|s|, |t| \leq R$.

Theorem 1.

(i) If $n=1$, then φ operates on IH^1 iff φ is locally Lipschitz. φ operates boundedly, in the sense that

$$\|(\varphi \circ g)'\|_{H^1} \leq C\|g'\|_{H^1} \tag{1.4}$$

for some $C<\infty$ and all $g \in IH^1$, iff φ is Lipschitz.

(ii) If $n \geq 2$, then φ operates on IH^1 iff φ is Lipschitz. Every such φ operates boundedly.

Remark. Let IH_0^1 be $\{\int_{-\infty}^{x} f(t)dt : f \in H^1\}$ when $n=1$, and $\{If : f \in H^1\}$ in general. Then IH_0^1 is the set of all $g \in IH^1$ which vanish at infinity in a suitable sense (e.g. the averages $|B|^{-1}\chi_B * g(x) \to 0$ as $|x| \to \infty$, for a fixed ball B; when $n=1$ we may simply require $g(x) \to 0$). It follows as in the proofs below that φ operates (boundedly) on IH_0^1 iff φ is locally Lipschitz or Lipschitz as in Theorem 1, and furthermore $\varphi(0) = 0$.

The proof of Theorem 1 depends on the following characterization of IH^1, which is interesting in its own right.

Theorem 2. Define the maximal operator N by

$$N(g)(x) = \sup_{t>0} t^{-n-1} \int_{|y-x|<t} |g(y) - g(x)|dy . \tag{1.5}$$

Then $g \in IH^1$ iff $N(g) \in L^1$, and $\|\nabla g\|_{H^1}$ and $\|N(g)\|_{L^1}$ are equivalent.

Maximal operators of the type (1.5) were introduced by Calderón [1] and further studied by Calderón and Scott [2] and DeVore and Sharpley [4]. (In particular, $N(g)$ as defined above equals a.e. $N_1^1(g)$ as defined in [2].) These references contain e.g. an analogue to Theorem 2 for Sobolev spaces.

We will prove Theorem 2 in Section 2 and then Theorem 1 in Section 3. Two final remarks are given in Section 4.

Acknowledgement. These results were obtained during the Seminar on Harmonic Analysis and Partial Differential Equations in El Escorial in June 1987 as an answer to the following question, raised by Joan Verdera in connection with spectral synthesis on H^1 and circulated among the participants of the seminar: Does (in our notation) f real and $f \in IH^1$ imply that $\max(f,0) \in IH^1$? (By Theorem 1, the answer is affirmative.)
I thank the organizers of the Seminar for their hospitality, and several participants, in particular Guido Weiss, for helpful discussions.

2. Proof of Theorem 2.

Lemma 1. Let φ be a test function with support in $\{x : |x| \le 1\}$ and let $A = \sup|\nabla\varphi|$. If $\varphi_t(x) = t^{-n}\varphi(x/t)$, then

$$\sup_{t>0}|\varphi_t * D_j g|(x) \le AN(g)(x) \quad , \quad j=1,\ldots,n \ . \tag{2.1}$$

Proof.

$$\varphi_t * D_j g(x) = D_j\varphi_t * g(x) = \int t^{-n-1} D_j\varphi\left(\frac{x-y}{t}\right)g(y)dy =$$

$$= t^{-n-1}\int D_j\varphi\left(\frac{x-y}{t}\right)(g(y) - g(x))dy \ ,$$

because $\int D_j\varphi = 0$. Hence

$$|\varphi_t * D_j g(x)| \le At^{-n-1}\int_{|y-x|<t}|g(y) - g(x)|dy \ ,$$

and (2.1) follows by the definition of $N(g)$. □

By choosing a φ as in the Lemma and integrating (2.1) we obtain, for some $C < \infty$,

$$\|D_j g\|_{H^1} \le C\|N(g)\|_{L^1} \ , \quad j=1,\ldots,n \ . \tag{2.2}$$

This proves the implication $N(g) \in L^1 \Rightarrow g \in IH^1$ in Theorem 2. The converse follows from the next lemma.

Lemma 2. Let $f \in H^1(\mathbb{R}^n)$ and define g by $g(x) = \int_{-\infty}^{x} f(t)dt$ when $n=1$ and $g = If$ when $n \ge 2$. Then,

$$\|N(g)\|_{L^1} \le C\|f\|_{H^1} \ . \tag{2.3}$$

Proof. Consider first the case $n=1$. It is clearly sufficient to prove (2.3) when f is an atom in H^1, and, by translation invariance, we may assume that, for some $R > 0$,

$$f \text{ is supported on } [0,R] \ ; \tag{2.4}$$

$$|f| \le 1/R \ ; \tag{2.5}$$

$$\int f dx = 0 \ . \tag{2.6}$$

It follows that g too is supported on $[0,R]$,

$|g| \le \|f\|_{L^1} \le 1$, and

$$|g(x) - g(y)| = |\int_x^y f(t)dt| \leq |x-y|/R .$$

Consequently, for every x ,

$$Ng(x) \leq \sup_t t^{-2} \int_{|x-y|<t} \frac{|x-y|}{R} dy = R^{-1} . \tag{2.7}$$

Furthermore, let $|x| > 2R$. Then

$$\int_{|x-y|<t} |g(y) - g(x)|dy = \int_{|x-y|<t} |g(y)|dy \leq \int_{-\infty}^{\infty} |g(y)|dy \leq R .$$

Since $g(y) - g(x) = 0$ when $|x-y| < t < |x|/2$, it follows that

$$Ng(x) \leq \sup_{t>|x|/2} t^{-2}R = 4R|x|^{-2} , \quad |x| > 2R \tag{2.8}$$

(2.7) and (2.8) now yield the desired estimate

$$\int_{-\infty}^{\infty} Ng(x) \leq 2 \int_0^{2R} R^{-1}dx + 2 \int_{2R}^{\infty} 4R\, x^{-2}dx = 8 .$$

In the case $n \geq 2$ we also use atomic decomposition, and may assume that

$$f(x) = 0 \quad \text{when} \quad |x| > R ; \tag{2.9}$$

$$|f(x)| \leq R^{-n} ; \tag{2.10}$$

$$\int f(x) = 0 . \tag{2.11}$$

Then

$$\int_{|y-x|<t} |g(y) - g(x)|dy = \int_{S^{n-1}} \int_0^t |g(x+s\xi) - g(x)|s^{n-1}ds\, d\xi \leq$$

$$\leq \int_{S^{n-1}} \int_0^t \int_0^s |\nabla g(x+u\xi)|s^{n-1}du\, ds\, d\xi \leq$$

$$\leq t^n \int_{S^{n-1}} \int_0^t |\nabla g(x+u\xi)|du\, d\xi = t^n \int_{|z|<t} |\nabla g(x+z)||z|^{1-n}dz \leq$$

$$\leq t^n \int_{|z|<t} |z|^{1-n}dz\, M(\nabla g)(x) = Ct^{n+1} M(\nabla g)(x) ,$$

where M is the Hardy-Littlewood maximal function. (The last but one step follows because $|z|^{1-n}$ is decreasing radial, of [5,III.2.2].) Consequently, since $D_j g = D_j If = R_j f$,

$$N(g)(x) \leq C\, M(\nabla g) \leq C \sum_1^n M(R_j f) \tag{2.12}$$

Since M and R_j are bounded on L^2 , we obtain

$$\int_{|x|<4R} N(g)(x)dx \le C R^{n/2}\|N(g)\|_{L^2} \le C R^{n/2}\Sigma\|M(R_j f)\|_{L^2} \le$$

$$\le C R^{n/2}\|f\|_{L^2} \le C . \qquad (2.13)$$

It remains to estimate $N(g)(x)$ when x is large. We begin by observing that $g(x) = C\int|x-y|^{1-n}f(y)dy$ [5,V.1.1]. We thus obtain, by (2.9)-(2.11),

$$|g(x)| = C\left|\int_{|y|<R} (|x-y|^{1-n} - |x|^{1-n})f(y)dy\right| \le$$

$$\le C R|x|^{-n} , \text{ when } |x| > 2R , \qquad (2.14)$$

Similarly,

$$|\nabla g(x)| \le C R|x|^{-n-1} , \quad |x| > 2R . \qquad (2.15)$$

Now suppose $|x| > 4R$. If $|y-x| < t < |x|/2$, then by (2.15) ,

$$|g(y) - g(x)| \le C t R|x|^{-n-1} .$$

Hence,

$$t^{-n-1} \int_{|y-x|<t} |g(y) - g(x)|dy \le C R|x|^{-n-1} , \quad t < |x|/2 . \qquad (2.16)$$

Furthermore, if χ_{2R} is the characteristic function of the ball $\{x : |x| < 2R\}$, then

$$\int_{|y|<2R} |g(y)|dy = \chi_{2R} * g(0) \le C \chi_{2R} * |x|^{1-n} * |f|(0) \le$$

$$\le C\|\chi_{2R} * |x|^{1-n}\|_{L^\infty}\|f\|_{L^1} \le C R . \qquad (2.17)$$

If $t \ge |x|/2$, we obtain by (2.14) and (2.17)

$$t^{-n-1} \int_{|y-x|<t} |g(y) - g(x)|dy \le t^{-n-1}\int_{|y|<3t} |g(y)|dy + Ct^{-1}|g(x)| \le$$

$$< t^{-n-1} \int_{|y|<2R} |g(y)|dy + t^{-n-1} \int_{2R<|y|<3t} C R|y|^{-n}dy + Ct^{-1}R|x|^{-n} \le$$

$$\le C Rt^{-n-1} + C Rt^{-n-1} \log\frac{3t}{2R} + C Rt^{-1}|x|^{-n} \le$$

$$\le C R|x|^{-n-1} \log\frac{|x|}{R} . \qquad (2.18)$$

Together with (2.16), this shows that

$$N(g)(x) \leq C R |x|^{-n-1} \log \frac{|x|}{R}, \quad |x| > 4R, \tag{2.19}$$

which yields

$$\int_{|x|>4R} N(g)(x)dx \leq C. \tag{2.20}$$

Lemma 2, and thus Theorem 2, follows by (2.13) and (2.20). □

3. Proof of Theorem 1.

It is obvious that if φ is Lipschitz then $N(\varphi \circ g) \leq C N(g)$, and thus φ operates boundedly on IH^1 by Theorem 2. Similarly, if $n=1$ then every $g \in IH^1$ is bounded, and thus $N(\varphi \circ g) \leq C N(g)$ (with C depending on $\|g\|_\infty$), as soon as φ is locally Lipschitz.

For the converse implications we consider the cases $n=1$ and $n \geq 2$ separately.

First let $n=1$. If I is any interval, let h_I denote the continuous function which is 0 outside I, 1 on $\frac{1}{2}I$, the middle half of I, and linear on the remaining two parts of I. Note that $h_I' \in H^1$ with norm independent of I ($\frac{1}{4}h_I'$ is a standard H^1-atom).

Suppose that φ is not locally Lipschitz. Then there exist $M < \infty$ and sequences $\{s_k\}$ and $\{t_k\}$ with $|s_k| \leq M$, $|t_k| \leq M$ and $|\varphi(s_k) - \varphi(t_k)| > 2^k |s_k - t_k|$. If $|s_k - t_k| > 2^{-k}$, then we subdivide the interval $[s_k, t_k]$ into a finite number N of intervals of length $|s_k - t_k|/N \leq 2^{-k}$. Since at least one of these subintervals, $[s_k', t_k']$ say, satisfies $|\varphi(s_k') - \varphi(t_k')| \geq |\varphi(s_k) - \varphi(t_k)|/N > 2|s_k' - t_k'|$, we may replace s_k, t_k by s_k', t_k' and thus assume that $|s_k - t_k| \leq 2^{-k}$. Furthermore, since the sequence $\{s_k\}$ is bounded, we may, by selecting subsequences, assume that $\{s_k\}$ converges to some s, and that $|s_k - s| < 2^{-k}$.

Let I be any interval, let $\{I(k)\}_{k=1}^{\infty}$ be disjoint intervals contained in $\frac{1}{2}I$, and let $\{I(k,j)\}_{j=1}^{N_k}$ be disjoint intervals contained in $\frac{1}{2}I(k)$, where $N_k = [2^{-k}/|s_k - t_k|]$. We then define

$$g = s h_I + \sum_{k=1}^{\infty} (s_k - s) h_{I(k)} + \sum_{k=1}^{\infty} \sum_{j=1}^{N_k} (t_k - s_k) h_{I(k,j)} .$$

By our construction

$$\|g'\|_{H^1} \le C|s| + \sum_{k=1}^{\infty} C|s_k - s| + \sum_{k=1}^{\infty} \sum_{j=1}^{N_k} C|t_k - s_k| \le$$

$$\le C|s| + C\sum_k 2^{-k} + \sum_k N_k|t_k - s_k| < \infty .$$

Thus $g \in IH^1$. On the other hand, $\varphi \circ g$ goes from $\varphi(s_k)$ to $\varphi(t_k)$ and back again on each interval $I(k,j)$. Hence

$$\|(\varphi \circ g)'\|_{H^1} \ge \|(\varphi \circ g)'\|_{L^1} \ge \sum_{k,j} \int_{I(k,j)} |(\varphi \circ g)'| \ge$$

$$\ge \sum_k N_k \cdot 2|\varphi(s_k) - \varphi(t_k)| \ge \sum_k 2N_k \cdot 2^k|s_k - t_k| = \infty ,$$

because $2N_k \ge N_k + 1 > 2^{-k}/|s_k - t_k|$. Consequently, $\varphi \circ g \notin IH^1$ and φ does not operate on IH^1.

If φ operates boundedly on IH^1, n still being 1, we let I and $I(k)$ be as above and define, for given s and t and a positive integer N,

$$g = sh_I + \sum_1^N (t-s)h_{I(k)} .$$

Thus $\|g'\|_{H^1} \le C|s| + CN|t-s|$, while, as above, $\|(\varphi \circ g)'\|_{H^1} \ge$ $\ge 2N|\varphi(t) - \varphi(s)|$. Consequently

$$|\varphi(t) - \varphi(s)| \le C\|g'\|_{H^1}/2N \le C|s|/N + C|t-s| ,$$

with C independent of s,t and N. Letting $N \to \infty$ we see that φ is Lipschitz.

When $n \ge 2$, the argument is slightly different. If B is any ball in R^n, let h_B denote the function which is 0 outside B, 1 on $\frac{1}{2}B$, and linear along the radii on the remainder of B. If B has radius r, then ∇h_B is bounded by $2/r$ and has support in the ball B with measure cr^n. Since the integral of ∇h_B obviously vanishes, this implies that $\|\nabla h_B\|_{H^1} \le Cr^{n-1}$.

Suppose that φ is not Lipschitz. Then there exist $\{s_k\}$ and $\{t_k\}$ with $|\varphi(s_k) - \varphi(t_k)| > 2^k|s_k - t_k|$. Let $\{B(k)\}_1^{\infty}$ be disjoint balls with radii r_k such that $|s_k|r_k^{n-1} \le 2^{-k}$. Let N_k be a large integer and let $\{B(k,j)\}_{j=1}^{N_k}$ be disjoint balls in $\frac{1}{2}B(k)$ with radii ρ_k such that $\rho_k^{n-1} = N_k^{-1} 2^{-k}|s_k - t_k|^{-1}$.

This construction requires that $N_k \rho_k^n$ be small enough (less than a constant depending on the dimension times r_k^n), but since

$$N_k \rho_k^n = 2^{-k} |s_k - t_k|^{-1} \rho_k = (2^k |s_k - t_k|)^{-n/(n-1)} N_k^{-1/(n-1)} ,$$

we can always achieve this by taking N_k large enough.

Let

$$g = \sum_{k=1}^{\infty} s_k h_{B(k)} + \sum_{k=1}^{\infty} \sum_{j=1}^{N_k} (t_k - s_k) h_{B(k,j)} .$$

Then $\|\nabla g\|_{H^1} \leq \Sigma\, C s_k r_k^{n-1} + \Sigma\, C N_k |t_k - s_k| \rho_k^{n-1} < \infty$ so $g \in IH^1$. On the other hand, if x_{kj} is the centre of $B(k,j)$,

$$\|\nabla(\varphi \circ g)\|_{L^1(B(k,j))} \geq \int_{S^{n-1}} \int_{\rho_k/2}^{\rho_k} \left|\frac{\partial(\varphi \circ g)}{\partial r}(x_{kj} + rw)\right| r^{n-1} dr\, dw \geq$$

$$\geq c\, \rho_k^{n-1} \int_{S^{n-1}} \int_{\rho_k/2}^{\rho_k} \left|\frac{\partial(\varphi \circ g)}{\partial r}(x_{kj} + rw)\right| dr\, dw \geq c\, \rho_k^{n-1} |\varphi(s_k) - (t_k)| ,$$

and thus

$$\|\nabla(\varphi \circ g)\|_{L^1} \geq \sum_{k=1}^{\infty} c N_k \rho_k^{n-1} 2^k |s_k - t_k| = \infty .$$

Thus $\varphi \circ g \notin IH^1$ and φ does not operate on IH^1. □

4. Further remarks.

1. Theorem 1 (i) holds also for the space $IH^1(\mathbb{T})$ of functions on the unit circle. This has the following application, pointed out by R.R. Coifman.

Consider the operators (defined e.g. on $L^p(\mathbb{R}^2)$, $1 < p < \infty$) of the form $Tf = cf + K * f$, where K is a Calderón-Zygmund kernel on $\mathbb{R}^2$ such that the restriction of K to the unit circle belongs to $H^1(\mathbb{T})$. Then the set of Fourier multipliers corresponding to these operators equals the set of functions on R^2 that are homogeneous of degree 0 and belong to IH^1 on $\mathbb{T}$. These operators form an algebra, which thus is isomorphic to $IH^1(\mathbb{T})$ [3,p. 600-601]. Theorem 1 (i) shows that any Lipschitz function (and no other) operates on this algebra of operators.

Note that a vector space of functions is an algebra (under pointwise multiplication) iff x^2 operates on the space. Hence Theorem 1 also gives a new prooof that $IH^1(\mathbb{R})$ and $IH^1(\mathbb{T})$ are algebras. Similarly, $L^\infty \cap IH^1(\mathbb{R}^n)$ is an algebra for every n.

2. Let us return to the original question with $\varphi(x) = \max(x,0)$, and consider real functions on $\mathbb{R}$. If $f \in H^1$ and $g' = f$ then the set where g is negative is a union of disjoint intervals; Theorem 1 says that if we redefine g to be 0 in these intervals, or equivalently redefine f to be 0 in these intervals, then the new f also belongs to H^1 . Boris Korenblum pointed out that this is not always true if we redefine f and g only in some of the intervals. For an example, let h be an odd function in H^1 which is supported on $[-1,1]$ such that $h(x) \leq 0$ for $x > 0$ and $\int_0^1 h = -1$, but $h \notin L \log L$. Let f equal h on $(-1,1)$, 1 on $(1,2)$, -1 on $(-2,-1)$, and 0 elsewhere. Then $\{x : g(x) = \int_{-\infty}^{x} f(t)dt < 0\} = (-2,0) \cup (0,2)$, but we do not obtain a function in H^1 by redefining f to vanish on one of these intervals.

<u>References</u>

1. A.P. Calderón, Estimates for singular integral operators in terms of maximal functions. Studia Math. 44 (1972), 563-582.

2. A.P. Calderón and R. Scott, Sobolev type inequalities for $p > 0$. Studia Math. 62 (1978), 75-92.

3. R.R. Coifman and G. Weiss, Extensions of Hardy spaces and their use in analysis. Bull. Amer. Math. Soc. 83 (1977), 569-645.

4. R.A. DeVore and R.C. Sharpley, Maximal functions measuring smoothness. Mem. Amer. Math. Soc. 293 (1984).

5. E.M. Stein, Singular integrals and differentiability properties of functions. Princeton Univ. Press, Princeton 1970.

6. H. Triebel, Theory of function spaces. Birkhäuser, Basel Boston Stuttgart 1983.

SPECTRAL SYNTHESIS IN SPACES OF FUNCTIONS WITH DERIVATIVES IN H^1

JOAN OROBITG

1. INTRODUCTION

We wish to look at functions with first partial derivatives in $H^1(R^n)$, the usual Fefferman-Stein Hardy space [3]. A convenient way of doing that is to consider the space

$$IH^1(R^n) = \{f : f = I_1 * h, h \in H^1(R^n)\}$$

where I_1 is the Riesz potential of order 1, defined by

$$I_1(x) = |x|^{1-n}, \qquad \text{if } n > 1,$$
$$I_1(x) = \log|x|, \qquad \text{if } n = 1.$$

We endow $IH^1(R^n)$ with the Banach space norm $\|f\| = \|h\|_{H^1}$, where $f = I_1 * h$. By the fractional integration theorem (e.g.[6]), functions in IH^1 belong to $L^{n/(n-1)}$. Thus

$$(1) \qquad \frac{1}{|B|}\chi_B * f(x) \longrightarrow 0 \qquad as \quad x \to \infty, \quad \text{for any fixed ball } B,$$

which can be thought of as a vanishing condition at infinity.

Our space IH^1 can be described also as the set of locally integrable functions f with derivatives in H^1 and satisfying (1).

Functions in $IH^1(R^n)$ are obviously continuous for $n = 1$, but this is no longer true for $n > 1$. It turns out that the right capacity measuring the deviation from continuity of functions in $IH^1(R^n)$ is the $(n-1)$- dimensional Hausdorff content M^{n-1}. Using the inequality (Adams [1])

$$(2) \qquad M^{n-1}(\{x : M(I_1 * h)(x) > \lambda\}) \leq \lambda^{-1}\|h\|_{H^1}$$

where M is the Hardy-Littlewood maximal operator, one can show that any $f \in IH^1$ has a (essentially unique) M^{n-1}-quasicontinuous representative. We say that a function g is M^{n-1}-quasicontinuous if, given $\epsilon > 0$, there exists an open set G with $M^{n-1}(G) < \epsilon$ such that g restricted to the complement of G is continuous there.

Fix now a closed subset F of R^n and ask under what conditions a function f can be approximated in IH^1 by functions in $C_o^\infty(F^c)$. It is not difficult to show that

$$(3) \qquad f = 0 \quad M^{n-1}\text{-a.e. on } F$$

is necessary. In this note we prove that (3) is also sufficient.

THEOREM. *Let $F \subset R^n$ be closed, let f be a M^{n-1}- quasicontinuous function in IH^1 satisfying (3). Then there exists a sequence (f_j) in $C_o^\infty(F^c)$ such that $f_j \longrightarrow f$ in IH^1.*

The proof of the theorem is based on a recent result of Janson [5] characterizing the space IH^1 by means of a suitable maximal operator.

As a final remark, let's mention that for $n = 1$ our result gives a characterization of closed ideals of the Banach algebra $IH^1(R)$.

As usual, we will write $A \sim B$ whenever the quantities A and B have ratios bounded by dimensional constants.

Research supported in part by the grant PB85-0374 of the CAICYT, Ministerio de Educación y Ciencia, Spain.

2. Proof of the theorem

We start with a brief discussion of Janson's result. For a locally integrable function f define the maximal operator

$$Nf(x) = \sup_{t>0} t^{-(n+1)} \int_{|x-y|<t} |f(x) - f(y)|\, dy$$

We have the following.

THEOREM [JANSON]. *A locally integrable function f belongs to IH^1 if and only if $Nf \in L^1$ and f satisties (1). Moreover, $\|f\| \sim \|Nf\|_1$.*

Consequently Lipschitz functions vanishing at the origin operate boun dedly on IH^1. In particular, IH^1 is stable under truncation, that is, if $f \in IH^1$ then $max(f,0) \in IH^1$. Since truncation is available, we can now follow the classical pattern of the proof of the spectral synthesis theorem for the Sobolev space $W_1^p(R^n)$ (Bagby [2], Hedberg [4]). We start by observing that is enough to prove the theorem for non-negative functions.

Given a non-negative $f \in IH^1$, $\epsilon > 0$ and $k > 0$, we define $f_\epsilon(x) = f(x) - \epsilon$ if $f(x) > \epsilon$ and $f_\epsilon(x) = 0$ otherwise, $f^k(x) = k$ if $f(x) > k$ and $f^k(x) = f(x)$ otherwise.

Part (ii) of the following lemma will allow us to prove the theorem just for (non-negative) bounded functions, and part (i) will be used later on.

LEMMA 1.

(i) $f_\epsilon \longrightarrow f$ *in* IH^1 *as* $\epsilon \to 0$.
(ii) $f^k \longrightarrow f$ *in* IH^1 *as* $k \to \infty$.

PROOF: We prove only (i) because (ii) follows similarly. Since $f - f_\epsilon = f^\epsilon$ we must show that $f^\epsilon \to 0$ in IH^1 as $\epsilon \to 0$. This is equivalent to $\partial_j f^\epsilon \to 0$ in H^1 as $\epsilon \to 0$, $1 \le j \le n$, because $\|f^\epsilon\| \sim \sum_{j=1}^{n} \|\partial_j f^\epsilon\|_{H^1}$.

Let φ be a test function with support in the unit ball, radially decreasing and with $\int \varphi = 1$. Set $\varphi_t(x) = t^{-n}\varphi(\frac{x}{t})$, $t > 0$, and $S^\epsilon(x) = \sup_{t>0} |\varphi_t * \partial_j f^\epsilon(x)|$.

Since $\|\partial_j f^\epsilon\|_{H^1} \sim \|S^\epsilon\|_1$, [3], it sufficies to show that

(4)
$$S^\epsilon \longrightarrow 0 \quad \text{in } L^1 \text{ as } \epsilon \to 0.$$

We have $S^\epsilon(x) \le M(\partial_j f^\epsilon)(x)$, ([7], p.62), and $\partial_j f^\epsilon = \partial_j f \chi_{\{f \le \epsilon\}} \to 0$ in L^1 as $\epsilon \to 0$. Hence, for any given sequence $\epsilon_n \to 0$ as $n \to \infty$, $S^{\epsilon_n} \to 0$ in measure, owing to the weak (1,1) inequality for the Hardy-Littlewood maximal operator. Therefore, passing to a subsequence, we get $S^{\epsilon_n} \to 0$ a.e. as $n \to \infty$.

By lemma 1 of [5],

$$S^\epsilon(x) \le \|\nabla\varphi\|_\infty N f^\epsilon(x)$$

and obviously $Nf^\epsilon(x) \le Nf(x)$.

Thus (4) follows from dominated convergence because $Nf \in L^1$. ∎

We come now to the proof of the theorem. Let f be as in the statement of the theorem, and assume, without loss of generality, that $0 \le f \le k$.

The M^{n-1}-quasicontinuity of f implies that $U_j = \{x : f(x) < j^{-1}\}$ is M^{n-1}-quasiopen. By this we mean that for each $\epsilon > 0$ there exists an open set Ω with $M^{n-1}(\Omega) < \epsilon$ and such that $U_j \cup \Omega$ is open. Consider now the set function

$$R_1(E) = \inf\{\|h\|_{H^1} : h \in H^1 \text{ and } I_1 * h \geq \chi_E\}.$$

This is a slight modification of a set function introduced in [1]. It turns out that $M^{n-1}(E) \sim R_1(E)$ for Borel sets E, but we will just show the inequality we need, namely, $R_1(E) \leq cM^{n-1}(E)$. This follows from $R_1(A) \leq R_1(B)$ if $A \subset B$, $R_1(\cup A_j) \leq \sum R_1(A_j)$ for arbitrary sets A_j, and the fact that the homogeneity of R_1 is $n-1$. The proof of the above three properties of R_1 is left to the reader. Let now $E \subset \cup Q_j$, where the Q_j are cubes with sides paral.lel to the coordinates axis, of length δ_j. Then

$$R_1(E) \leq \sum R_1(Q_j) = c \sum \delta_j^{n-1},$$

and so $\quad R_1(E) \leq cM^{n-1}(E)$.

Therefore, given a positive integer j we can find an open set Ω_j with $R_1(\Omega_j) < j^{-1}$ and such that $U_j \cup \Omega_j$ is open. Observe that $U_j \cup \Omega_j$ contains F because U_j does.

Define $\psi_j = \max(f - j^{-1}, 0)$. Then $\psi_j = 0$ on U_j, $\|\psi_j\| \leq c\|f\|$ and $\|\psi_j\|_\infty \leq k$. The definition of R_1 gives us a function $g_j \in H^1$ with $I_1 * g_j \geq \chi_{\Omega_j}$ and $\|g_j\|_{H^1} < j^{-1}$. Put

$$\omega_j = \min(1, I_1 * g_j),$$

so that $\omega_j = 1$ on Ω_j, $0 \leq \omega_j \leq 1$, $\|\omega_j\| < j^{-1}$ and ω_j is M^{n-1}-quasicontinuous.

Set $f_j = \psi_j - \psi_j\omega_j$. Then $f_j = 0$ M^{n-1}-a.e. on the neighbourhood $U_j \cup \Omega_j$ of F. We claim that $f_j \longrightarrow f$ in IH^1 as $j \to \infty$.

We have

$$\|f - f_j\| \leq \|f - \psi_j\| + \|\psi_j\omega_j\| = \mathbf{I}_j + \mathbf{II}_j.$$

Lemma 1 (i) says that $\mathbf{I}_j \to 0$ as $j \to \infty$, so we only need to worry about $\mathbf{II}_j$.

Since $\|\psi_j\|_\infty \leq k$ and $N\psi_j \leq Nf$, we get

$$N(\psi_j\omega_j)(x) \leq \omega_j(x)Nf(x) + kN\omega_j(x),$$

and then

$$\mathbf{II}_j \sim \|N(\psi_j\omega_j)\|_1 \leq \|\omega_j Nf\|_1 + k\|N\omega_j\|_1.$$

Now $\|N\omega_j\|_1 \sim \|\omega_j\| \to 0$ as $j \to \infty$, and so we are left with the task of showing that $\omega_j Nf$ tends to 0 in L^1.

By (2) $\quad M^{n-1}(\{x : \omega_j(x) > \lambda\}) \leq c(\lambda j)^{-1}$.

Hence, passing to a subsequence, which we again denote by ω_j, we obtain $\omega_j \to 0$ a.e. as $j \to \infty$, and then $\omega_j Nf \to 0$ a.e. as $j \to \infty$.

The above convergence turns out to be dominated because $\omega_j Nf \leq Nf \in L^1$.

The proof of the claim is now complete.

We have got approximants f_j with support contained in F^c. We need now to modify them to get compact support in F^c, and this is what the next lemma accomplishes. After that, a standard regularization argument will provide us with the desired approximants in $C_o^\infty(F^c)$.

LEMMA 2. *For each positive integer k let ω_k be a function in $C_o^\infty(R^n)$ such that $0 \le \omega_k \le 1$, $\omega_k(x) = 1$ if $|x| \le k$, $\omega_k(x) = 0$ if $|x| > 2k$ and $|\nabla\omega_k| \le const\, k^{-1}$.*
Then $f\omega_k \longrightarrow f$ in IH^1 as $k \to \infty$, for each $f \in IH^1$.

PROOF: Set $\beta_k = 1 - \omega_k$. We have to show that $N(f\beta_k) \to 0$ in L^1 as $k \to \infty$.
Clearly

$$N(f\beta_k)(x) \le \beta_k(x)Nf(x) + \sup_{t>0} t^{-n-1} \int_{|x-y|<t} |f(x)||\beta_k(x) - \beta_k(y)|\, dx$$
$$\equiv P_k(x) + Q_k(x),$$

and so

$$\|N(f\beta_k)\|_1 \le \|P_k\|_1 + \|Q_k\|_1.$$

Since β_k vanishes on $|x| < k$, $\|P_k\|_1 \le \int_{|x|>k} Nf(x)\, dx$, which tends to zero as $k \to \infty$ because $Nf \in L^1$.
We claim now that

$$\|Q_k\|_1 \le const\, k^{-1} \int_{|x|<3k} Mf(x)\, dx \tag{5}$$

Let's first show how one concludes the argument once (5) is established. Since $Mf \in L^{n/(n-1)}$, because $f \in L^{n/(n-1)}$, we only need to prove that for each $g \in L^{n/(n-1)}$ one has

$$k^{-1} \int_{|x|<k} |g(x)|\, dx \longrightarrow 0 \qquad as \quad k \to \infty.$$

This is clear for $g \in C_o^\infty$, and the general case follows from

$$k^{-1} \int_{|x|<k} |g(x)|\, dx \le \|g\|_{n/(n-1)},$$

which is an obvious consequence of Holder's inequality.
To prove (5) we will use the following pointwise estimates for Q_k:

$$Q_k(x) \le const\, k^{-1} Mf(x) \qquad , x \in R^n, \tag{6}$$

$$Q_k(x) \le (|x| - 2k)^{-n-1} \int_{|y|<2k} |f(y)|\, dy \quad , |x| > 2k. \tag{7}$$

The gradient estimate for β_k gives (6) readily, and (7) is an easy consequence of fact that $\beta_k = 1$ on $|x| > 2k$. We have now

$$\|Q_k\|_1 = \int_{|x|<3k} Q_k(x)\, dx + \int_{|x|>3k} Q_k(x)\, dx = \mathbf{I} + \mathbf{II}.$$

To estimate **I** one uses (6) to get

$$\mathbf{I} \le const\, k^{-1} \int_{|x|<3k} Mf(x)\, dx\,,$$

and using (7) **II** is estimated by

$$\int_{|y|<2k} |f(y)|\,dy \int_{|x|>3k} (|x|-2k)^{-n-1}\,dx \le ck^{-1} \int_{|y|<2k} |f(y)|\,dy$$
$$\le ck^{-1} \int_{|y|<3k} Mf(y)\,dy.$$

The proof of (5) is complete. ∎

Acknowledgement. I am grateful to Joan Verdera for posing the problem.

REFERENCES

1. Adams,D.R., *A note on the Choquet integrals with respect to Hausdorff capacity*, Proceedings of the US-Swedish Conference on Function Spaces and Interpolation Theory (1986), Lund, Sweden.
2. Bagby,T., *Quasi topologies and rational approximation*, J. Functional Analysis **10** (1972), 259-268.
3. Fefferman,C.L., Stein,E.M., *H^p spaces of several variables*, Acta Math. **129** (1972), 137-193.
4. Hedberg,L.I., *Non-linear potentials and approximation in the mean by analitic functions*, Math. Z. **129** (1972), 299-319.
5. Janson,S., *On functions with derivatives in H^1*, these proceedings
6. Krantz,S.G., *Fractional integration on Hardy spaces*, Studia Math. **73** (1982), 87-94.
7. Stein,E.M., "Singular integrals and differentiability properties of functions," Princeton Univ. Press, 1970.

Universitat Autònoma de Barcelona
Departament de Matemàtiques
08193 Bellaterra (Spain)

An Averaging Operator on a Tree

Richard Rochberg and Mitchell Taibleson

I. Introduction: We obtain estimates for some operators acting on functions defined on a homogenous tree. Our main result is a weak type (1,1) estimate for a certain averaging operator.

Our interest in these operators derives from the role they, and closely related operators, play in the harmonic analysis of functions on free groups, on martingales, and on $\mathbb{R}^n$. However, we will not pursue those connections here, rather we indicate briefly in the final section where to find out more about them.

Our approach here is essentially elementary and combinatoric. We find it intriguing that we obtain our weak type result without use of a covering lemma. The key idea is that in order to show that an operator with positive kernel is of weak type (1,1) it suffices to show that its formal adjoint is of type (∞,∞) on a rich class of characteristic functions. (See Proposition 3 and the proof of Theorem 4.) This approach to weak type (1,1) estimates was described by Peter Jones at the conference reported in these Proceedings.

II. A Weak–Type Estimate: X is a tree, homogeneous of degree three, in its half–plane realization. We think of the tree as a collection of nodes; the paths that connect nearest neighbors are thought of as a relation on the tree. For $x \in X$ we identify exactly two *lower neighbors* of x and a unique *upper neighbor.* Between any two nodes in X there is a unique shortest path that connects the two nodes; this path is called a *geodesic.* A node x is said to have a *descendent* in a set A if there is an $a \in A$, $a \neq x$, and the geodesic $\{x = x_0, x_1, \ldots, x_n = a\}$ has the property that x_k is a lower neighbor of x_{k-1}, $k = 1,2,\ldots,n$. A node x is said to have a descendent in a set A that *derives from* a lower neighbor y if y is in A or if y has a descendent in A. Suppose $E \subset X$. The *lower boundary* of E, ∂E, is the subset of E consisting of those nodes which have no descendents or whose descendents all derive from just one of its lower neighbors.

This research was supported in part by NSF Grants DMS–8701271 and DMS–8701263.

NOTATION. $|E|$ denotes the cardinality of the set E.

LEMMA 1. *If* E *is a finite subset of* E *then* $|\partial E| > \frac{1}{2}|E|$.

PROOF. B, the "bottom of E", denotes those points of E that have no descendents in E; $T = E \backslash \partial E$, is the "top of E". For each node $x \in X$ x, along with all its descendents, forms an infinite triangle with x at the vertex. Such a triangle can be truncated below to form a "finite full triangle" Δ. Suppose that E is contained in such a triangle where the vertex of Δ is a point of E. Let G be the graph obtained by drawing all geodesics that connect the vertex of Δ to the other points of E. Let Br be the branch points of G. It is easy to see that

$$|B| = |Br| + 1.$$

It is not difficult to check that $x \in T$ if and only if $x \in E \cap Br$, so $|T| \leq |Br|$. It is trivial that $B \subset \partial E$ so $|B| \leq |\partial E|$. Consequently,

$$|\partial E| \geq |T| + 1,$$

and since ∂E and T partition E, it follows that $|\partial E| > \frac{1}{2}|E|$. This completes the proof if E is contained in such a triangle Δ.

If E is a finite subset of the tree then we may decompose E, $E = \bigcup_{k=1}^{n} E_k$, so that for each k, $E_k \subset \Delta_k$, where Δ_k is a full finite triangle and some point of E_k is the vertex of Δ_k. This can be done so that the infinite extensions of the triangles are pairwise disjoint. To get the decomposition take a highest node in E, consider the triangle with vertex at that node and then cut back to a minimal finite triangle that does not exclude points of E. This gives Δ_1 and we set $E_1 = E \cap \Delta_1$. Now take a highest node in $E \backslash E_1$ and construct Δ_2 and E_2 in the same manner. In a finite number of steps we construct the required decomposition. It is clear that $\partial E = \bigcup_{k=1}^{n} \partial E_k$. Since $|\partial E_k| > \frac{1}{2}|E_k|$ for each k it follows that $|\partial E| > \frac{1}{2}|E|$. □

We define two positive real operators. For $x \in X$ consider the infinite geodesic $\{x_0, x_1, x_2, \ldots\}$ where $x_0 = x$, and x_{k+1} is the upper neighbor of x_k for $k = 0,1,2,\ldots$. For f a function on the tree we set

$$Tf(x) = \sum_{k=0}^{\infty} 2^{-k} f(x_k).$$

We label the nodes in the infinite triangle with its upper vertex at x as follows: $x_{01} = x$; x_{11}, x_{12}

are the lower neighbors of x_{01}. Similarly $x_{k1},\dots,x_{k2^k}$ are the 2^k lower neighbors of $x_{k-1,1},\dots x_{k-1,2^{k-1}}$. We set

$$Sf(x) = \sum_{k=0}^{\infty} \sum_{j=1}^{2^k} 2^{-k} f(x_{kj}).$$

Formally $T^* = S$ as operators on L^2(X,counting measure).

LEMMA 2. *If* E *is a finite subset of the tree* X *then* $S\chi_{\partial E}(x) < 2$ *for all* x.

PROOF. Observe that if x is a node of the tree and y and z are the two lower neighbors of x then

$$(*) \qquad Sf(x) = f(x) + \tfrac{1}{2}\Big[Sf(y) + Sf(z)\Big].$$

Set $f = \chi_{\partial E}$. Clearly if $x \notin E$ and x has no decendents in E then $Sf(x) = 0$. If $\{E_k\}$ is the decomposition of E given in Lemma 1, and $\{\Delta_k\}$ is the corresponding collection of triangles, then we only need to consider x if $x \in \Delta_k$ for some k or if x has a descendent in some Δ_k.

Fix a k. If x is a node in the lowest level of Δ_k then $Sf(x) = f(x)$ which is equal to 0 or 1 and so it is less than 2. We argue by induction on the level of x in E_k. Suppose we have shown that $Sf(y) < 2$ for all y in E_k below the level of x. There are two possibilities: 1)If $x \notin \partial E$ then from $(*)$ we have that $Sf(x) < 0 + \frac{1}{2}(2+2) = 2$. 2)If $x \in \partial E$ let y and z be the lower neighbors of x. Since $x \in \partial E$ at least one of y and z is not in E and has no descendents in E, say y. Then $f(x) = 1$, $Sf(y) = 0$, and $Sf(z) < 2$, so from $(*)$ we have that $Sf(x) < 1 + \frac{1}{2}(0+2) = 2$. This completes the argument if x is in some Δ_k. The case that remains is when x is in none of the triangles but has a descendent in one or more of the triangles. Then one argues by induction on the level. If x is such a point then $f(x) = 0$. By the induction hypothesis $Sf(y)$ and $Sf(z)$ are less than 2, where y and z are the lower neighbors of x, so by $(*)$ again, we see that $Sf(x) < 2$. This completes the proof of the second lemma. □

PROPOSITION. 3. *If* E *is a finite subset of the tree there is a subset* F *of* E, *such that* $|F| > \frac{1}{2}|E|$ *and* $S\chi_F(x) < 2$ *for all* x.

This is an immediate consequence of Lemma 1 and Lemma 2.

THEOREM 4. *If* $f \in L^1(X)$ *and* $s > 0$ *then*

$$|\{x: |Tf(x)| > s\}| \le \frac{4}{s} \|f\|_1.$$

PROOF. Since $|Sf(x)| < S|f|(x)$ we may assume that $f(x) \ge 0$ for all x. We may also assume that $\|f\|_1 = 1$. It is easy to see that if $f \in L^1$ then f is bounded and tends to zero at infinity. From this it follows that if $s > 0$ then $E = \{x: Tf(x) > s\}$ is finite. Now choose F as in Proposition 3. By Fubini's theorem

$$\frac{1}{2}|E| \cdot s < \int Tf\, \chi_F = \int f\, S\chi_F < 2.$$

The theorem follows. □

III. An Interpolation Result: We will say that a function f on X is in the space CM_d if Sf is bounded. The letters in the name are for *"discrete Carleson Measure"*. Regard X in the usual way as the index set for the dyadic intervals of $\mathbb{R}$. For x in X let c(x) and l(x) denote the center and length of the corresponding dyadic interval. f is in CM_d exactly if the measure on $\mathbb{R}^2_+$ given by

$$\Sigma\, f(x)\, \delta_{(c(x),l(x))}\, l(x)$$

is a Carleson measure. Clearly,

$$CM_d \subset \ell^\infty(X). \tag{3.1}$$

Using the results of the previous section we will show that CM_d can be used as a substitute for ℓ^∞ as an end point for interpolation. We refer to [BL] for the notation and terminology of interpolation theory. In particular, we recall that a natural endpoint of the scale of spaces $\ell^p(X)$ is $\ell^0(X)$, the set of functions on X with finite support. We measure size in that space by

$$\|f\|_0 = |\mathrm{supp}(f)|.$$

The result we will generalize is that the real interpolation spaces between ℓ^p and ℓ^∞ are the Lebesgue–Lorentz spaces $\ell^{r,q}$. That is, for $0 \le p < \infty$, $0 < \theta < 1$, $1 \le q \le \infty$, and $r = p/(1-\theta)$

$$(\ell^p, \ell^\infty)_{\theta,q} = \ell^{r,q}. \tag{3.2}$$

THEOREM 5. *For* $0 \le p < \infty$, $0 < \theta < 1$, $1 \le q \le \infty$, *and* $r = p/(1-\theta)$

$$(\ell^p, CM_d)_{\theta,q} = (\ell^p, \ell^\infty)_{\theta,q} = \ell^{r,q}.$$

PROOF. By reiteration it suffices to consider the case $p = 0$. (3.2) is the right equality. The inclusion of the space on the left in the center space follows from (3.1). To obtain the other inclusion we will work at the K functional level and follow a pattern of proof used by Sagher [S] in generalizing results of Bennet, DeVore and Sharpley.

For a function a defined on X we recall the definition of the K functional for the couple (ℓ^0, CM_d). For $t > 0$

$$K(t,a) = K(t,a;\ell^0,CM_d) = \inf\{\|a_0\|_0 + t\|a_1\|_{CM_d} : a_0 + a_1 = a\}.$$

We will show that there is a $c > 0$ so that for $0 < t \le s$

$$K(t,a) \le c(\tfrac{t}{s}\|a\|_0 + t\|a\|_\infty(1 + \log\tfrac{s}{t})), \tag{3.3}$$

and, for $t > s > 0$

$$K(t,a) \le c(\|a\|_0(1 + \log\tfrac{t}{s}) + s\|a\|_\infty). \tag{3.4}$$

Although (3.3) and (3.4) are not the hypothesis used in Theorem 2 of [S], they are exactly the two estimates needed to make that proof work. Thus, when we have these estimates we can appeal to the proof in [S] to conclude that the identity map is bounded from $(\ell^p,\ell^\infty)_{\theta,q}$ to $(\ell^p, CM_d)_{\theta,q}$ and we will be done.

Suppose $t \simeq 2^{-k}s$ for a positive integer k. Let E be the support of A. Set $\partial_1 = \partial E$. Inductively, for $j = 1,2,\dots,k$, set

$$\partial_{j+1} = \partial(E \setminus \bigcup_{i<j} \partial_i).$$

For $j = 1,2,\dots,k$ set $b_j = a\,\chi_{\partial_j}$. Set $a_1 = \sum_{j=0}^{k} b_j$ and $a_0 = a - a_1$.

$$\begin{aligned} K(t,a) &\le \|a_0\|_0 + t\|a_1\|_{CM_d} \\ &\le |E \setminus \bigcup_{i=1}^{k} \partial_i| + t\|\Sigma b_j\|_{CM_d} \\ &\le |E \setminus \bigcup_{i=1}^{k} \partial_i| + t\,\Sigma\|b_j\|_{CM_d} \\ &\le |E \setminus \bigcup_{i=1}^{k} \partial_i| + t\,\Sigma\|b_j\|_\infty \|\chi_{\partial_j}\|_{CM_d}. \end{aligned}$$

By Lemma 1, $|E \setminus \bigcup_{i=1}^{k} \partial_i| \le 2^{-k}|E|$. Certainly, $\|b_j\|_\infty \le \|a\|_\infty$. By Lemma 2, $\|\chi_{\partial_j}\|_{CM_d} = \|S\chi_{\partial_j}\|_\infty \le 2$. Combining these facts we continue the estimate with :

$$\le 2^{-k}|E| + 2kt\,\|a\|_\infty.$$

This is (3.3) for these s,t. The case of general s,t follows with a factor c. (3.4) is elementary because it is always true that $K(t,a) \le \|a\|_0$ and, if $t > s$, this is less than the right and side of (3.4). □

COROLLARY 6. *For* $1 < p < \infty$, *the operators* S *and* T *are bounded maps of* ℓ^p *to* ℓ^p.

PROOF. By Theorem 4, T is weak type (1,1). It is easily seen to be of strong type (∞,∞). The boundedness follows from the Marcinkiewicz interpolation theorem. For S, the strong type (1,1) estimate is elementary and the boundedness from CM_d to ℓ^∞ is by definition; the boundedness of ℓ^p follows from Theorem 5. □

(In fact S and T are adjoints and either half of the corollary implies the other. Also, since both operators have positive kernels, direct proofs of the boundedness can be constructed easily.)

IV. Remarks: The space CM_d and the operator S are used is a series of papers [RS1,2,3] which study singular integral operators on $\mathbb{R}^n$. In particular, Theorem 5 extends a result in [RS1] and Theorem 4 can be used to give an alternative proof of one of the results in [RS3].

The operator T also shows up in studying the relationship between harmonic functions defined on the tree X (i.e., functions which satisfy the natural mean value property) and dyadic martingales. More precisely, a function on the tree is said to be *regular* or *harmonic* if its value at each node is the average of its values at its three nearest neighbors, while a dyadic martingale may be viewed as a function on the tree whose value at each node is the average of its two lower nearest neighbors. It has been shown that the operator T gives a one–to–one onto correspondence between the class of martingales and the class of harmonic functions (under trivial normalizing conditions) [KPT]. The theorems above on the boundedness of T on ℓ^p can be reformulated as results about spaces defined by certain mixed norms that characterize Besov–Lipschitz and Triebel–Lizorkin spaces, and the equivalence of these characterizations in terms of boundary martingales and harmonic extensions. Examples of such norm characterizations for Besov spaces in terms of martingales can be found in [OS].

REFERENCES

[BL] J. Berg and J. Lofstrom, *Interpolation Spaces, An Introduction*, Springer Verlag, 1976.

[KPT] A. Koranyi, M. Picardello, and M. Taibleson, *Hardy spaces on non–homogeneous trees*, to appear in Springer Verlag Lecture Notes, Proc. Conf. Harmonic Anal., Cortona, Italy, 1984.

[OS] C. W. Onneweer and W.–Y. Su, *Homogeneous Besov spaces on locally compact Vilenkin groups*, to appear in Studia Mathematica.

[RS1] R. Rochberg and S. Semmes, *A decomposition theorem for functions in BMO and applications*, J. Functional Anal, 67, (1986) 228–263.

[RS2] ____________________, *Nearly weakly orthonormal sequences, singular value estimates*, and Calderon–Zygmund operators, Manuscript, 1987.

[RS2] ____________________, *End point results for estimates of singular values of singular integral operators*. Manuscript, 1987.

[S] Y. Sagher, *A new interpolation theorem*, Harmonic Analysis, Proceedings, Minneapolis 1981, Lecture Notres in Math. 908, 189–198.

Washington University – Saint Louis, MO 63130